BIOPSYCHOLOGY

THE BASICS

Biopsychology: The Basics is a concise, accessible and illuminating introduction to the field of biopsychology.

The book explores what psychology is in the broadest sense and how combining it with a biological perspective offers a deeper understanding of behavior and mental life. Key topics include the following:

- What biopsychology is: understanding the interaction of biology and psychology
- The biology of the brain and how to study it
- How psychological states are related to physiological processes
- The effects of drugs, both therapeutic and recreational, on behavior and psychology
- How genes and the environment impact psychological development
- The biopsychology of cognition
- People in the world: understanding emotions, motivation and communication
- The biological basis of psychopathologies – causes, diagnoses and treatments
- Explanations, mechanisms and the biopsychology of consciousness

With suggestions for further reading and an extensive glossary of key terms, this book is an engaging and ideal introduction for those coming to the subject for the first time.

Philip Winn is Emeritus Professor of Neuroscience at the University of Strathclyde, Scotland, United Kingdom.

Madeleine Grealy is Professor of Psychology at the University of Strathclyde, Scotland, United Kingdom.

The Basics Series

The Basics is a highly successful series of accessible guidebooks which provide an overview of the fundamental principles of a subject area in a jargon-free and undaunting format.

Intended for students approaching a subject for the first time, the books both introduce the essentials of a subject and provide an ideal springboard for further study. With over 50 titles spanning subjects from artificial intelligence (AI) to women's studies, The Basics are an ideal starting point for students seeking to understand a subject area.

Each text comes with recommendations for further study and gradually introduces the complexities and nuances within a subject.

BIOPSYCHOLOGY
PHILIP WINN AND MADELEINE GREALY

ISLAMIC PSYCHOLOGY
G. HUSSEIN RASSOOL

IMITATION
NAOMI VAN BERGEN, ALLARD R. FEDDES, LIESBETH MANN AND BERTJAN DOOSJE

SELF AND IDENTITY
MEGAN E. BIRNEY

PSYCHOPATHY
SANDIE TAYLOR AND LANCE WORKMAN

EVOLUTIONARY PSYCHOLOGY
WILL READER AND LANCE WORKMAN

WORK PSYCHOLOGY
LAURA DEAN AND FRAN COUSANS

FORENSIC PSYCHOLOGY (3RD EDITION)
SANDIE TAYLOR

APHASIA
JANE MARSHALL

For more information about this series, please visit: www.routledge.com/Routledge-The-Basics-Series/book-series/B

BIOPSYCHOLOGY
THE BASICS

Philip Winn and Madeleine Grealy

LONDON AND NEW YORK

Cover image: Nobi_Prizue via Getty Images

First published 2025
by Routledge
4 Park Square, Milton Park, Abingdon, Oxon OX14 4RN

and by Routledge
605 Third Avenue, New York, NY 10158

Routledge is an imprint of the Taylor & Francis Group, an informa business

© 2025 Philip Winn and Madeleine Grealy

The right of Philip Winn and Madeleine Grealy to be identified as authors of this work has been asserted in accordance with sections 77 and 78 of the Copyright, Designs and Patents Act 1988.

All rights reserved. No part of this book may be reprinted or reproduced or utilised in any form or by any electronic, mechanical, or other means, now known or hereafter invented, including photocopying and recording, or in any information storage or retrieval system, without permission in writing from the publishers.

Trademark notice: Product or corporate names may be trademarks or registered trademarks, and are used only for identification and explanation without intent to infringe.

British Library Cataloguing-in-Publication Data
A catalogue record for this book is available from the British Library

ISBN: 978-1-032-10474-4 (hbk)
ISBN: 978-1-032-10472-0 (pbk)
ISBN: 978-1-003-21550-9 (ebk)

DOI: 10.4324/9781003215509

Typeset in Bembo
by SPi Technologies India Pvt Ltd (Straive)

CONTENTS

List of illustrations — vi
Acknowledgments — viii

Introduction: What is biopsychology? — 1
1 The brain and how to get at it — 16
2 Physiology for psychologists — 46
3 The chemistry of the brain: Psychopharmacology — 60
4 Development — 80
5 Cognition — 101
6 Emotions, motivation and communication — 122
7 Psychopathologies — 141
8 Overall, what does biopsychology offer? — 162

Glossary — 168
Further reading — 186
References — 189
Index — 192

ILLUSTRATIONS

FIGURES

1.1 The basic structure of a neuron. Neurons vary greatly in their architecture. For example, the extent of their dendrites, whether or not axons are myelinated, the numbers of branches an axon has. All of these vary, but the basics – dendrites, cell body, axon, terminals – are constant   18

1.2 The basic structure of a synapse. The processes involved in synaptic transmission are complex and are further outlined in Chapter 3   19

1.3 The spinal cord as it passes down the spinal column. The letter s and numbers are indices of position. The letter C stands for Cervical, T for Thoracic, L for Lumbar, S for Sacral and Coc for Coccygeal. The cross-section shows the relative positions of areas of spinal cord gray matter. The dorsal horn is also known as the posterior horn; likewise, the ventral horn is also known as the anterior horn. The Rexed laminae (not shown) run up and down the length of the spinal cord. In the cross-section here, Lamina 1 is at the tip of the dorsal horn, with the subsequent laminae running through to the bottom of the anterior horn   24

1.4 A cross-section of the brain showing the positions of some key subcortical structures – the basal ganglia and the thalamus   32

1.5 An external view of the human brain showing the major cortical lobes   37

1.6 This illustrates the brain as if it had been cut in half along the long axis. It makes visible lobes of the cortex and key subcortical structures   39

2.1 The human heart. As we make clear, what happens in the body has a significant influence on biopsychological activity. Changes in heart rate are associated with a variety of cognitive and emotional states   47

2.2 The human immune system, which is being increasingly recognized as important in biopsychology. The illustration shows how the presence of the immune system in the brain, where the glymphatic system, which drains into the lymphatic system of the body, has a key role in maintaining brain health   49

TABLES

1.1 The Rexed laminae of the spinal cord   26
1.2 The cranial nerves   29
3.1 Common pharmaceutical treatments for psychiatric and neurological disorders and their effects on brain neurotransmitter systems   77
4.1 The structure of sleep   96
7.1 Conditions typically considered under the heading neurodiversity   143
7.2 The diagnostic criteria and codes of *DSM-5-TR*   148

ACKNOWLEDGMENTS

This book is meant for a wide readership. It's a starting point for anyone curious about biopsychology and doesn't require any prior knowledge or experience. As authors, we have tried to bring to bear our own knowledge of research and experience of teaching diverse groups of students from many different academic backgrounds – psychology, biology, neuroscience, engineering, pharmacy and others. We've tried hard to keep the text accessible, engaging and stimulating and gratefully acknowledge the assistance of several people in preparing the book. Jane Winn read each chapter in draft, advising on points of clarity and style. Susan Tyler prepared all the illustrations. Molly Selby, our editor at Routledge, guided us through the process of preparing the book. We are very grateful to each of them for their thoughtful and professional assistance.

INTRODUCTION

WHAT IS BIOPSYCHOLOGY?

What are people like? What sort of properties make a person? What makes us unique individuals? Are we different from other animals only by degree, at a separate point on a continuum, or are we a different kind of animal altogether? These are big questions that all sorts of scholars using all sorts of approaches try to address, and biopsychologists are in the mix with them. So what is biopsychology, and how does it help us to address these questions? We'll go at this step-by-step, starting with vocabulary.

The word 'biopsychology' seems to have two academic subjects running into each other, like biochemistry does, creating a borderland subject apparently cutting across the territory of two others. You might guess from the way the word 'biopsychology' is constructed that psychology, the fully spelled out part, is primary, with the bio- prefix indicating that we're going to be looking at how biological ideas and theories can inform what psychologists do. But then you look and see that other words covering exactly the same sort of material as we'll find in this book reverse biopsychology into psychobiology or expand it to biological psychology. And then you make the discovery, first, that the preferred term used to be physiological psychology and, second, that there's a whole slew of descriptions for an assortment of ever more specialized research fields: psychophysiology, psychopharmacology, comparative psychology, evolutionary psychology, both behavioral and cognitive neuroscience, as well as psychoimmunology, psychoendocrinology and more. You can permutate the prefixes psycho-, neuro- and elements of the life sciences to your heart's content. All of these

DOI: 10.4324/9781003215509-1

overlapping genres point, with differing degrees of specificity, at relationships between what we can think of as broadly biological and psychology, which inevitably asks an obvious backdrop question: what is psychology? Once we have a reasonable idea of what we mean by that, then we can think about what kind of relationships might exist between it and all the various life sciences, including the challenging idea that psychology is really no more than biology by another means. We need to think about these so that we have some degree of clarity when we begin to look inside biopsychology.

A VERY SHORT HISTORICAL REVIEW

What is psychology? It's a question asked countless times by innumerable people and – spoiler alert – there isn't an absolutely fixed way to define it, any more than there is any other academic subject. Academic disciplines don't actually ring-fence in the way that school examinations or university degrees seem to suggest. One of our colleagues, an eminent professor of chemistry, becomes quite agitated at the suggestion that the discovery of DNA was a triumph of biology. Not so! It's a triumph of chemistry! Of course, it doesn't greatly matter because the study of DNA and the mechanisms of genetic inheritance necessarily involve both chemistry and biology (and more). Academic subjects overlap like circles in a Venn diagram, and it can be hard to find the central defining point that identifies most clearly what a particular subject is all about. For psychology, we can think of its core business in two related ways, neither of them totally unique to the subject. Its principal concerns have to do with behavior and the mind, including all the processes that go with it – perception and attention, emotion and motivation, language, consciousness, cognition and so on. The mind and behavior seem obvious fundamentals for psychology, don't they? For this basic introduction, we're going to stick to this fundamental idea, but there are many philosophers and psychologists who are uncomfortable with the idea of minds, mental life and consciousness – too vague, too ephemeral, not physical enough. Thinking about the mind is a philosophical exercise that goes back millennia and seems set to remain a controversial topic forever. Much more surprisingly, Daniel Levitis and his colleagues at UC Berkeley reported in a scientific paper that biologists could not agree on what constitutes behavior. They arrived at a definition: "behavior is the internally coordinated responses (actions or inactions) of whole living organisms (individuals or groups) to internal and/or external stimuli, excluding responses more easily understood as developmental changes". It makes sense – behavior as what an animal chooses to do or not to do – but it seems at best rather labored and at worst borderline absurd to be so concerned about something that people talk about every day without needing to worry about definition. We'll proceed on the assumption that readers will have a sensible enough idea of what we mean by behavior and mental life.

Because psychology is a subject with a clear interest in the human condition, it's no surprise that cultures worldwide pay attention to it. For example, both Confucianism and Buddhism speak to aspects of psychology. However, given that our principal concern is with biopsychology as a scientific discipline – and the fact that *psyche* derives from Greek – it's appropriate to start with how ancient Western philosophy dealt with it.

Aristotle was the author of *Peri Psyches* or, in Latin, *De Anima*. Both Greek and Latin translate this as "about the soul", although variations have included "about the mind" or "about the psyche". In his book, Aristotle summarized the work of previous Greek philosophers and talked about such things as the senses, memory and personality, all in a theoretical and philosophical way rather than as experimental science. Central to his thinking was the idea of a soul. Aristotle's idea of the soul is a little alien to modern psychological sensibility. What he meant by it was not something distinct from the body but the essence of that body. Souls of specific kinds were what made their owners what they were, giving essence to their material bodies. The idea of a soul has continued to develop in Western thought. The biggest jump was when St. Thomas Aquinas (1222–1274) brought Christian theology to bear on Aristotelian thinking. Scholars, theologians and philosophers continue to argue about the existence and nature of souls. Such discussion is not something that is found in most psychology books or courses, and indeed, most biologists and psychologists treat the notion of a soul with cold hard skepticism. Nevertheless, despite his terminology not equating to that of modern science, Aristotle did lay the ground for later approaches to psychology. He believed that everything reduced to the operation of matter, and he believed that all knowledge comes from experience, that we begin at birth as a blank slate, a *tabula rasa*, which experience writes on to create us as persons – and experience of course is inherently bound up with processes that psychologists are very interested in, among them perception, learning and memory: experience as what we detect happening in and around us and what we do with that information.

You find a distant shadow in ancient Greek and Roman thought of something important to contemporary biopsychology. Have you ever wondered why our everyday talk about emotions is expressed in terms of body systems? Things like being all heart, getting your blood up, not having the stomach for something or the possession of guts, as bravery, instincts or feelings. It comes from the Greek medic Hippocrates who, around 400 BC, created a theory in which different body fluids – he called them humors – had different effects on health and emotional life. Blood, black bile, yellow bile and phlegm, each contained in a different part of the body and associated with different conditions. We still use the words: to be sanguine is to be optimistic and positive – sanguine is to do with blood; to be choleric is to be bad tempered – yellow bile; to be melancholic is to be sad – black bile; and to be phlegmatic is to be unemotional, caused by phlegm. The Roman physician Galen in the second century of the Christian era added more, arguing that the quality of food – hot or cold, wet or dry – would affect the humors and push a person's psychology in one direction or another. It's easy to treat all this with a benign yet patronizing tolerance, but we'll see relationships between brains, body systems and psychology reminiscent of what Hippocrates was on about. Not being on the money with your methods doesn't mean your principles can't be sound.

It would be all but impossible to minimize the impact of ancient Greek philosophy on Western thought. It weaves right through to the Middle Ages, only very gradually being challenged by the development of a more objective scientific method during the Renaissance in the 1500s and in the Enlightenment period that followed in the 1700s. For instance, Andreas Vesalius (1514–1564) published *De Humani Corporis Fabrica Libri Septem* (Seven Books on the Structure of the Human Body), breaking with the until-then acknowledged master, Galen, to lay the foundations for a modern approach to anatomy and the physical structure of the human body. Vesalius was followed by William Harvey (1578–1657), who described the circulation of blood, and Thomas Willis (1621–1675), who looked at the brain, discriminating for the first time between its gray and white matter as well as describing its blood supply. He's memorialized by the Circle of Willis, an arrangement of arterial blood flow at the base of the brain. Acceleration of progress would follow through the 1800s. For examples, we can look at three events, occurring close to each other in time, that predicted how psychology would develop.

First event: in 1796, David Kinnebrook was sacked from his post as assistant to Astronomer Royal Nevil Maskelyne. Examining the transit of stars across the night sky by observing through a telescope and using the then standard "eye and ear" method – looking while counting off time – Kinnebrook was consistently 0.8 seconds out from Maskelyne. At that time, it was assumed that the eye and ear method was infallible and that any two observers would always get the same result. The fact that it didn't happen in this case was treated as a matter of hierarchy: the Astronomer Royal was right, his assistant wrong and consequently fired. However, Friedrich Bessel (1784–1846), an astronomer from Königsberg, grasped that this was a universal problem and that two observers of the same event would almost always produce consistently different readings. Bessel developed what was called the personal equation, a correction factor that attempted to eliminate interindividual differences. Fascinating, but why is this important? Because such work would become a central plank of psychophysics, the systematic attempt to describe human sensory abilities in terms of physical laws, creating mechanical accounts of sensation and perception. Physical laws would, it was hoped, put psychology on the same scientific footing as other major nineteenth-century sciences, like physics, chemistry and geology.

Second event: in 1798, a boy who for several years had been living a wild, feral existence in southern France was finally caught. He became known as the Wild Boy of Aveyron. He was taken to Paris where some hoped that he would be found to be a physical representation of the noble savage, a perfect man, uncorrupted. He wasn't: he was wild, unkempt and a distinctly unnoble savage, without language or emotion, showing little evidence of any intelligence or cognitive ability aside from the rudimentary skills that had kept him alive. Regarded by many physicians as a hopeless case, he came under the care of a young doctor, Jean Marc Gaspard Itard, who began a systematic series of tests and training. Itard was, in surprisingly short order, able to coax from Victor, as he named him, a degree of psychological performance that approached something near normality. Why is this important? Because it is the first well-documented case of systematic intervention, a clear program designed to enhance intelligent understanding and modify behavior, eliminating the inappropriate and crude and developing something better.

Last in this triple bill is Francis Gall (1758–1828). In 1801, his lectures were banned by Emperor Francis I of the Holy Roman Empire for being too radical, too dangerous and too disruptive. The reason why might seem strange. Francis Gall was the anatomist who created the science of phrenology. Science? Phrenology was the last word in scientific thinking at that time but is regarded now as pseudoscience: reading bumps on the skull to determine what went on below in the brain. The method is nonsense because the outer surface of the skull bears no relationship to the brain matter below. The inner surface of the skull does, but that wasn't where Gall was looking. What so disturbed the authorities wasn't so much the method but two core features of Gall's thinking: that we can account for all psychological processes through the operation of the brain and that the volume of a particular part of the brain will correlate to ability: more brain would mean better, faster, longer-lasting functions. It was this explicitly material account of people – no soul here – that was radical enough to bring the ban. All of this would be so much historical noise were it not for one thing: Gall's methods were wholly wrong; his description of psychological functions was wrong in both principle and terminology. There is, for example, no specific location in the brain to account for a sense of marvelousness, as was suggested. But the central idea of localization of function, this psychological process here in the brain, that one there, really truly stuck and has continued to inform biopsychology more-or-less to the present day. It's a problem we'll have to face up to later on in the book. For now, what's important about Francis Gall is the establishment of a biological account – a neurobiological account – of psychology.

Everything accelerated through the 1800s, major advances in biology and psychology, creating exciting, if sometimes wrongheaded, developments. Perhaps the pivotal moment came in 1879 when Wilhelm Wundt (1832–1920) established the Institute for Experimental Psychology at the University of Leipzig in Germany. The opening of this Institute is rightly credited as being the moment when psychology established itself as an independent discipline, applying a properly rigorous, scientific approach to the study of psychological and mental processes. There are any number of famous names to cite and discoveries that we could record, but for the purpose of this fleeting history, the most important thing to note is that virtually everyone was working in the belief that scientific laws could be established. Psychology was to stand with the other disciplines as a science of material things. The idea that psychological events have physical explanations persists.

Through much of the 1800s, there was an emphasis on human psychology. In parallel, biology was having a series of revolutions. For instance, with advances in microscopy, Theodor Schwann (1810–1882) introduced the idea that animals were made of cells: it's rather hard to imagine such a basic fact being discovered as opposed to having always been known. Claude Bernard (1813–1878) pioneered physiology, introducing the important concept of homeostasis, the tendency of cells, systems and whole organisms to maintain a constant state. (Claude Bernard didn't use the word 'homeostasis': the American physiologist Walter Cannon (1871–1945) was actually the first to do so.) Neurologists were busy exploring the brain, refining anatomy using new technologies to examine neural tissue in ever more detail. Nobel prize winner Santiago Ramon y Cajal (1852–1934) created neuron doctrine, the idea that the brain was made of individual cells, cells that came to be called neurons. But of course, dominating the biology of the 1800s was Charles Darwin and the theory of evolution.

The idea of a unity to the animal kingdom, all evolved from a far distant common ancestor, was a spur to scientists interested in animal behavior. Edward Thorndike (1874–1949) was one such psychologist, publishing 'Animal Intelligence' in 1898, but he was far from alone. The earliest stand-out individual was Ivan Petrovich Pavlov (1849–1936), a Nobel prize-winning Russian physiologist whose work created the idea of conditioning, dogs learning to associate the sound of a bell with the presentation of food, such that the bell alone would elicit physiological changes, salivation being the most visible.

The growing interest in testing the behavior of lab animals led inexorably to the behaviorist school of thought in the twentieth century, with Burrhus Frederic Skinner (1904–1990) – usually referred to as just B. F. Skinner – probably the best-known and most influential theorist. Behaviorists examined stimuli and responses in exquisite detail, with various schedules of reinforcement determining the relationships between them. What's so intriguing is the abrupt turn away from the 1800s investigations into mental processes. Twentieth-century behaviorists had no interest in mental life, believing it to be essentially introspective and beyond the reach of scientific analysis. For very many researchers, examination of visible, quantifiable behavioral events, not the unseen mental ones, became the primary point of psychology.

Inevitably, it broke. Behaviorism had its successes, the most significant being the development of clinical behavior modification techniques that have been of benefit to a great many people. However, several things came together in the mid-twentieth century that reined in behaviorism: a recognition of the serious difficulty behaviorist principles have in accounting for the incredible flexibility of human language; rapid advances in understanding the anatomy and physiology of the brain; the development of information theory, better computers and artificial intelligence; and the discovery – a rediscovery in many ways – that it was possible to take objective and quantifiable measurements of mental abilities. The fusion of these things created cognitive science, morphing into cognitive psychology and cognitive neuroscience, both of which position themselves to analyze the processes of psychology – sensation, perception and attention; learning and memory; decision-making; communication and socialization; motivation; and even emotions. All of these could be investigated as both independent and integrative processes, and each could, in theory at least, be tagged to localized brain activity using current psychological terminology and scientific technology.

So here we are in the present, still interested in behavior and still interested in the events and processes that make up our mental lives. Psychology has come a long way since Aristotle wrote *Peri Psyches*. Knowledge has expanded, and technologies have improved, but still we wrestle with the same problems of our psychological nature. Given that we have an idea of how psychology has developed and what it's trying to do, how do we imagine that having a bio-edge helps? What does biopsychology offer?

WHERE DOES BIOPSYCHOLOGY TAKE US?

Even from this briefest of histories, we can see that psychology traces roots that are at least two and a half millennia deep. Wanting to understand our status and nature as persons is not a new thing. Biopsychology has a similar depth to psychology *per se* – Hippocrates wanted to give a bodily basis to human states and conditions. However, depth of history doesn't at all imply a consistent rate of development. The authority of ancient philosophy went unchallenged from classical antiquity right through to the Middle Ages, after which the radical reformulations of the Renaissance moved the sciences, arts and humanities in bold new directions. The Renaissance was succeeded in turn by the Enlightenment, celebrating reason as the power that would allow people to understand the universe and people's place in it. Psychology is a term that would have been understood by Enlightenment thinkers, biopsychology not. Nevertheless, Bessel, Itard and Gall are all great examples of people who brought reason to bear in trying to understand how we work, how we can be fixed when we go wrong and who we are. They were, if you like, proto-biopsychologists. What really moves the subject on are the big ideas of the nineteenth century: the belief that psychology is a material subject of laws and that human beings are evolved animals. Accept these, and it seems that you inevitably buy into the idea that our physical structure – our biology – is implicitly bound up with our behavior and our mental lives and so, *voila*, biopsychology.

Even so, there is a very difficult question here. Talk of biopsychology leads many people immediately to assume that it ought to be possible to explain all of psychology – behavior, mental life, the whole thing – in terms of biology. Indeed, you'll find going through this book that we talk about nervous systems, the brain, spinal cord and all the nerves running through your body. We'll talk about the cardiovascular system, the endocrine and immune systems and even microbiology, looking at how gut bacteria influence our lives. It's a lot of biology having a lot of impact on our psychological lives, behavior and mind. Will it be sufficient to explain all of psychology? That's much less certain. To begin unpacking all of this, we need to talk about philosophy.

A commonly accepted principle in any sort of problem-solving is Occam's Razor: when trying to make sense of something, the explanation that requires the smallest number of assumptions is likely to be best – or more loosely, you might say 'keep it simple'. This is always attributed to William of Ockham (1287–1347) – Ockham: Occam is a misspelling – but in actual fact, the same principle can be found well before him in writings right back to Aristotle. And why razor? Disappointingly boring, it's just the commonplace idea of shaving away what's unnecessary. Nevertheless, despite getting the name wrong and the idea of razor being mundane, Occam's Razor is an important principle. The problem is where it can lead: reductionism.

Reductionism isn't one thing but several. To start, you can take a reductionist approach to solving problems. When your computer doesn't work, you try to revive it holistically by switching it off and switching it on again. If that doesn't solve the problem, you attempt to identify what specific thing might be wrong with it. This will involve a sort of constitutive reductionism, the method of analyzing a complex phenomenon by dissecting it stepwise into elements, shifts in the level of analysis from systems to components – simplifying reductive steps taken to make analysis possible. We all do this. Theory reductionism, completely collapsing one theory into a more basic one, such as a comprehensive theory of the brain couched in terms of physics, is more problematic. The same applies to explanatory reductionism, where the assumption is that knowledge of components will explain properties of whole systems. For example, if you understood all the properties of single nerve cells, you would be able to show how such cellular properties offered a necessary and sufficient explanation of a system property, like cognition or consciousness.

Some scientists make the reductionist assumption that a biological explanation of psychology is better because it's somehow deeper and more fundamental. That is, if you understood all there was to know about biology, you would inevitably understand everything there was to know about psychology as well. Neurobiologists are prone to this, believing that any process of psychological importance happens inside the skull and can be explained by the activity of nerve cells there. To disagree is to lay oneself open to the accusation of substance dualism – the belief that bodies and minds are separable things, different sorts of stuff – as if there were just two polar opposites to choose from. What we need to do is go a bit deeper and try for a more nuanced approach to biopsychology.

It's not common for people to accept substance dualism, and we are not going to take up that position. Like many other psychologists, we're comfortable with the philosophical idea of physicalism. Philosophers go into this in depth, but the essence of it is simple: there are physical explanations for everything, and there's nothing spooky required to describe things, nothing supernatural. Talking about physicalism and physical explanations isn't to say that physics as a subject explains everything. Physics is a subject that's good on the structure of things but mostly silent on the nature of things, which sounds odd. Think of it through this example: physics can tell us all about the structure of water, but what does it say about the quality of wetness? We could just say that wet means made of liquid or moisture, but that doesn't capture the quality of wetness: it just relabels liquid as wet, which isn't very satisfying. Materials scientists might say that wetness happens when a liquid contacts the surface of a solid, requiring cohesive and adhesive forces. But if wetness is a liquid sticking to a solid surface – like water on our skin – we can't say that water is wet *per se* because it takes both the liquid and the solid to define wetness. In a similar vein, when we treat wetness as a sensation we experience, then, fine, water is wet but only in so far as that is how we chose to describe that experience. We could dance around this all day long, but we won't. The point is this: that the structure and nature of things are not the same. Even so, it's still rational to believe that both are properties that can, now or at some point in the future, be explained without needing recourse to anything that goes beyond the natural world.

Why is all this important? Because, first, we are working on the belief that there are rational explanations for everything and that we don't need spooky stuff in order to account for things biological or things psychological. But that doesn't mean that we have to be reductionists, boiling psychology down into biology. There are two important considerations here. First, complexity – as in the interactions between the parts of things – is important because it creates properties that otherwise don't exist. Simple example: if you took a bicycle completely apart, every nut, every bolt, all the chain links, all the spokes, everything, what would you have? A pile of parts, none of which would have the property of a bicycle and none of which alone could predict the existence of such a thing. Bicycles are machines designed to transform human energy into more effective and efficient motion. They require assembly and, when properly put together, have properties that none of their individual parts have. Such properties are called emergent properties. Diving back into biology, cells do more than their parts, and collections of cells do more than individual cells can. Adopting a reductionist approach to psychology and biology can negate this complexity.

Second, we need to think a little more about what biologists get up to. We can put aside theories of evolution, not because they're unimportant but because there are other volumes dealing with both evolutionary biology and evolutionary psychology. Keeping a focus on biopsychology, there's an immediate and obvious tendency to think in terms of brains and bodies. Think back to Hippocrates and the four humors and their relation to emotional states, or think about modern neurobiology and investigations of the relationships between the brain and psychology. But biology is about more than the cells and organs of a body. For one thing, it has a big stake in behavior, as is apparent from the daily flood of natural history programs streaming through our TVs. There's something unexpected here for biopsychologists to do with behavior. Niche construction is the process by which an animal works to alter and shape its own environment, as well as the environments of others. There are examples from across the animal kingdom. The most obvious and visible is home building – birds build nests, rabbits dig warrens and so on. Then there's larger scale landscaping, of which beavers are a fine example: the dams they build create ponds and lakes that not only shape river environments but also change the local ecology. And on a smaller scale, earthworms modify the soil in which they live, changing it physically and chemically so that they can survive in it. Charles Darwin recognized the importance of earthworms in his 1881 book 'The Formation of Vegetable Mould Through the Action of Worms, with Observations on Their Habits'.

These examples aren't typical parts of a biopsychology program, so why do we think they're important? They are of serious interest to biopsychology because people create environments as well. We can characterize these as hard ones – the built environments we live in – and soft ones, the cultural environments we create and which shape our everyday lives. Biopsychology can accommodate a biological influence into psychology by more than just the parts that make bodies, us as physical beings. It can use biological theories that look at behavior as well, and in doing that, it can help move us on from a narrowly reductionist approach to one that can deal with persons in a more complete way, incorporating cells, systems, bodies, the environment and culture. And when we start thinking about the relationships between culture and psychology, we can for sure still see the influence of biology, but at the same time, we can grasp that an only-biological explanation of human culture might not be best. The description and analysis of human societies and cultures can identify biological roots through things like niche construction but will need a different sort of language to discuss what flourishes from them.

To wrap this up, as biopsychologists, do we have to assume that everything is ultimately reducible to biology? No. Can we talk about mental states that are not only or fully explained biologically? Yes. Does the bio-approach to psychology have value? Yes, definitely, but with a word of caution. History can help us see where ideas come from. We can celebrate important discoveries but also reflect on missteps that have been taken under the broad heading of biopsychology, whether people at the time used the expression or not. Here are two important examples.

First example: eugenics, which began in the late 1800s. Sir Francis Galton (1822–1911), a cousin of Charles Darwin, was the person who developed fingerprinting, creating a system of analysis still in use. A great achievement, but one overshadowed by the fact that he is also credited as the founder of eugenics. It was developed further by his student, Karl Pearson (1857–1936), a biostatistician who became the first Galton Chair of Eugenics at University College London. Eugenicists took an early idea of genetic inheritance (which we now know to be inadequate) and a naïve idea of social structure and behavior (which we are now rightly appalled by) and created eugenics, a subject promoted by eminent scientists and accepted by those in power. The central idea is explosively simple: the lower orders in society breed faster than their betters, and their genetics make them irredeemably dim-witted. Predicted outcome? Societies become swamped by the less able and in time will collapse. Actual effect of the theory? Sterilization programs to curb growth in the numbers of the less able or of groups less desirable. Such programs were widespread across the world throughout the twentieth century, some still just coming to light. Appalling.

Second example: brain surgery. António Caetano de Abreu Freire Egas Moniz (1874–1955), usually known as Egas Moniz, was a Portuguese neurologist who made important discoveries having to do with measuring blood flow in the brain – cerebral angiography as it's called. He was awarded the 1949 Nobel Prize for Physiology or Medicine, but not for this. The award was for his work in developing the prefrontal lobotomy as treatment for mental illness. Many medics and scientists were skeptical about the procedure at the time; many others adopted it with gusto. As time went by, psychiatry recognized that the technique was not having the desired effects, which, to be honest, is hardly surprising, and it was abandoned, though techniques of what is generically called psychosurgery persist.

We look at eugenics and lobotomies with horror. If we make a sober reflection on these events we might come to appreciate that those promoting them may have had good intentions. Or we might not: we might just see them as wrong, full stop. There is, though, a lesson to learn. Scientists and philosophers talk about "the provisional nature of knowledge", meaning that what we think we know right now may turn out to be incorrect. If we're going to apply scientific knowledge in the service of social policy or medicine, we need to be very confident that we're getting it right and not doing harm.

That note of caution sounded, what we'll see in the following chapters is what biopsychology offers across a range of topics that help us understand what we are as persons, how we develop, how we think, what our social and emotional lives are like and what happens when things go wrong. This last one is really important because as we're well aware, there is a crisis in mental illness, with too many people suffering and not enough effective diagnoses and treatments. We don't want to be glib about this, but it also highlights a simple point. If you're anxious or depressed or suffering in any similar way, who are you going to consult? A biologist or a psychologist? The answer is plain and will keep you focused on what's really important here. Biology has a lot to offer psychology, a lot that can be successfully integrated, but at the bottom line, whichever other subjects inform it, psychology is its own subject.

INTRODUCTION: SUMMARY

- *Psychologists study both behavior (those things that people and animals do) as well as the mind – processes like sensation, perception and attention; learning and memory; and emotion, motivation, language and consciousness.*
- *Biopsychology is the interaction between biology and psychology; it is not an attempt to reduce all psychological states and events to biological processes but a science that seeks to understand the interaction between the two.*
- *Regardless of what title it's given, biopsychology has a long history: trying to understand how biology influences psychology and vice versa is an ancient occupation. Though they wouldn't have described themselves as biopsychologists, it's possible to trace ideas back to ancient authors such as Hippocrates and Galen.*
- *While we don't seek to reduce psychology to biology, we assume that there are physical explanations for biological and psychological states and events. Equally, we recognize that complex states and events cannot necessarily be explained in terms of their elements – properties can emerge that are unpredictable.*
- *Biopsychology is a subject whose material is of great interest and potential practical relevance. We sound a note of caution about overinterpretation and premature use of new biopsychological discoveries, which may turn out to be less viable than anticipated.*

THE BRAIN AND HOW TO GET AT IT

Usually, when people think about biopsychology they think about brains, which is fair enough: the brain is important. However, focusing on the brain alone is insufficient. Brains are not things apart but parts of nervous systems. We separate out the central nervous system (CNS) – the brain and spinal cord – from the peripheral nervous system (PNS), all the nerves running through the body. But CNS and PNS are part of one package, both derived from the same developmental place. What's more, the nervous system throughout the body – brain, spinal cord, nerves – interacts with many other physiological systems, endocrine, immune, cardiovascular, digestive and so on. The next chapter will look at body physiology, but to start, let's look at the biology of the nervous system, how we study it and what some of the problematic issues are.

THE STRUCTURE OF THE NERVOUS SYSTEM: NEURONS AND NERVES

The origin of the nervous system illustrates its unity. It begins very early in human embryonic development with the formation of the neural plate, a sheet of cells. This folds over to create the neural tube, filled with cerebrospinal fluid (CSF), a mildly salty solution. The center of the neural tube develops into four ventricles – two lateral, one on the right and one on the left, plus the third and fourth along the middle – all of which are linked and are continuous with the central canal of the spinal cord. In this extended space, CSF is continually produced so that it can bathe all the cells in the

CNS, as well as operate as a drainage system. In humans, the neural tube is in place by the fourth week of pregnancy, after which it inexorably grows: the brain develops from the most forward part of the tube, with the remainder becoming the spinal cord. The PNS develops from the neural crest on top of the neural tube: one nervous system from one neural plate. (We'll look in more detail at this in Chapter 4.)

What about the cells of the nervous system? Let's be clear: neurons are individual nerve cells, nerves are not. Nerves are collections of neurons bound together by thin layers of connective tissue, like biological clingfilm. We'll talk about nerves in a moment when we look at the PNS. When human infants are born, virtually all the neurons they will ever have are present, some 86 billion. Present but not connected: what wires them together, and how does it happen?

One way or another, literally all the cells in the human body communicate by releasing chemicals to affect the activity of their immediate neighbors. What makes neurons special is not that they communicate *per se* but the specialized way in which they do it through synaptic transmission. Like any other cells, neurons can release chemicals indiscriminately from multiple points, but transmission at synapses is targeted, not indiscriminate.

The standard model of a neuron is simple and is illustrated in Figure 1.1. There is a cell body (aka soma) filled with a gel called cytoplasm in which critical functions are performed to keep the cell functioning properly. It contains small organelles, like the nucleus (from where genetic events are controlled) and mitochondria (which produce energy). Extending from the cell body are two kinds of extensions, dendrites and an axon, which are called processes (noun) because they process (verb) from the cell body. Dendrites are activated by inputs from other neurons. These activations are combined in the cell body, and the neuron will make an electrical response to them. Neuroscientists talk about the proximal and distal portions of dendrites, meaning the parts closest to the soma (proximal) and furthest away (distal): they can have different properties in coding the information that comes into a neuron. Many dendrites possess, sticking out more or less at right angles, little dendritic spines (which come in various types) where inputs from other neurons converge, with one of the inputs regulating

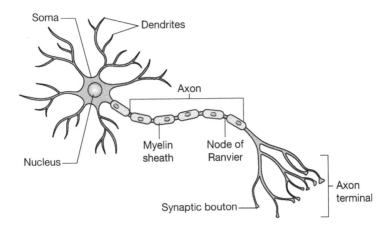

Figure 1.1 The basic structure of a neuron. Neurons vary greatly in their architecture. For example, the extent of their dendrites, whether or not axons are myelinated, the numbers of branches an axon has. All of these vary, but the basics – dendrites, cell body, axon, terminals – are constant.

what comes from the other. Those dendrites without spines are called aspiny. Dendrites are the principal targets for inputs from other neurons. Axons go the other way, carrying messages away from the cell body. Axons are slender tubes, given shape by microfilaments and microtubules running their length. As well as giving shape – a cellular skeleton – they also create separate channels. The cytoplasm in axons is renamed axoplasm and is in constant motion to and from the cell body, carrying chemicals from one point to another: channels are like the lanes on a highway, transport moving one way or another. Axons aren't like a wire with one end point but are branched with many endings. (For clarity, when people talk about fibers in the brain, they usually mean axons.) When electrical impulses get to the end of the axons, they arrive at synaptic junctions where the axon terminals make contact with the dendrites of other neurons. A synapse has structure – Figure 1.2 illustrates this. Miniature filaments hold it in place, connecting neuron-to-neuron (most often axon-to-dendrite; other arrangements are possible), but at its heart is a gap, the synaptic cleft. When electrical impulses arrive at synapses, they trigger the release of chemicals that cross

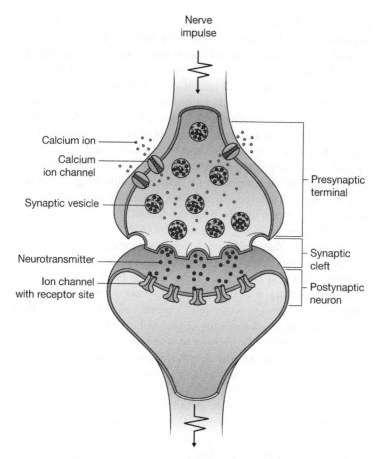

Figure 1.2 The basic structure of a synapse. The processes involved in synaptic transmission are complex and are further outlined in Chapter 3.

the cleft to activate receptors on the receiving neuron. Receptors work in a lock-and-key manner, with specific transmitters binding to specific receptor types. This is neurotransmission, the release of neurotransmitters across the synaptic cleft, a physical passing of information from one neuron to another.

Neurotransmission and its relationship to psychology and psychiatry is something we'll return to in Chapter 3 because brain chemistry and its management are significant parts of biopsychology.

For now, we'll concentrate on brain development and the creation of connectivity after one more bit of biology: glial cells.

Glial comes from Latin and means glue, which is what scientists originally thought glial cells were: glue holding the brain together. There are as many glial cells (aka glia) in the brain as there are neurons, but the neurons-to-glia ratio varies considerably over different parts. Are glia actually glue? Perhaps, in so far as they provide physical support for neurons, which couldn't exist without them, but the notion of glue fails to capture the diverse and critical work of glial cells. They come in different forms: star-shaped astrocytes provide support and nutrition to neurons, influencing everything they do; microglia scavenge debris and aid healing after damage; and oligodendroglia provide electrical insulation. (An aside: this sort of thing will come up a lot – different names for the same thing: astroglia/astrocytes; oligodendroglia/oligodendrocytes, where the suffix '-cytes' means cells. Microglia are always microglia, no substitutions.)

Where glial cells come from is interesting. Neural stem cells can divide to create more of themselves by a process called mitosis, two daughter cells formed from one original cell. However, as they divide, they create another type of glial cell called radial glia, which also divide. Through the division of radial glia, astrocytes and oligodendroglia are created. This does not happen with microglia; they are derived from blood and belong to the immune system. Radial glia also create neural progenitor cells (aka neuroblasts) that do not divide again and develop into neurons. So, rather than being glue, glial cells are critical for the development, support and health of neurons. The only thing they don't do is transmit information at synapses, though even here they play a role in enabling neurons to communicate.

Neurons are wired together through synaptic connections. How this happens draws in two components that resonate through psychology: genes and environment. The brain isn't made by all its parts growing independently and then clipping together like a model airplane kit. It starts with migration, where populations of neural progenitor cells and neurons move around, shepherded by radial glial cells. The migration is controlled by guidance signals in the form of proteins whose manufacture is genetically controlled. As neurons migrate to find an appropriate home, their processes develop. Axons grow away from cell bodies, sometimes only over a short distance, staying within the part of the brain where the cell body sits, but sometimes extending over longer distances, helping connect one structure to another. Their journeys are guided by growth cones at the tips of the axon, which sense chemical activity, luring them to the right place and keeping them away from the wrong one. When they arrive where they're supposed to, adhesion molecules on receptive neurons guide them to make contacts and synapses form. All this migration and movement is controlled by families of genes. It can be disrupted by environmental events such as dietary vitamin shortage or too much alcohol in the bloodstream via the placenta, leading to brain malformation in the developing embryo. In a healthy body, genes will guide the essentials of development in a way that is absolutely necessary. An essential beginning but not the whole story.

Genes guide the formation of the brain during early development, but after birth, the fine details are shaped by activity-dependent and experience-dependent development. You can think of these as being, in a sense, internal and external. Experience-dependent development often involves critical periods or sensitive periods of development. Critical ones are single opportunities that, if missed, will have bad consequences; sensitive ones are more flexible and recur – by analogy, miss a train, wait for the next one. Sensory systems offer examples of a critical period of development. A frequently cited one is binocular vision, which will not develop normally if a critical period of development is skipped, for instance, by being denied light. The important point here is that development requires more than just genetic activity: it needs the experience of what's happening in the world. Activity-dependent developments have more to do with the activity of the brain itself, including, for example, increased synapse formation when parts of the brain are particularly hard at work. Another activity-dependent development involves oligodendroglia. These provide a substance called myelin, which wraps around axons, insulating them. It might appear paradoxical, but insulation speeds up electrical transmission along axons. Rather than an electrical impulse traveling all along an axon, every micron of the way, in a myelinated fiber, the impulse can jump between gaps in the myelin sheath called the nodes of Ranvier. It makes the conductance of electrical impulses along axons orders of magnitude quicker. The assumption was always made that which neurons had myelin and which did not was genetically determined. To some extent, that's right, but not completely. There are degrees of myelination: increased use of a neuron ups the level of insulation. Activity-dependent and experience-dependent development mean that the wiring of the brain is not fixed (genetically determined) but adaptable in terms of both synaptic connection and speed of transmission.

Clearly, what happens to an infant during development will affect how its brain develops. But when does brain development stop? Human brain development mostly finishes at the end of adolescence. Axons have been guided into place, myelination has happened, synapses have formed – and then synaptic pruning begins. During development, the brain over-connects and forms more synapses than needed. Because they require so much energy, synapses that are not essential are eliminated (pruned), leaving a brain that's maximally effective and efficient. Is that the end of it? No, one last thing: adult neurogenesis. It used to be axiomatic that adult brains could not form new neurons, but this is wrong. There are some places in the brain – some places, not everywhere – in which new neurons can be created in adulthood. Not enough is known about this yet, but it intrigues many scientists because of the insight it might give into treatments for diseases in which neurons die prematurely.

THE STRUCTURE OF THE NERVOUS SYSTEM: THE SPINAL CORD

The spinal cord, with the brain, forms the CNS. It is one continuous entity. The ventricular system runs through it, and it's all surrounded by the meninges, which are three layers of connective tissue: the dura mater (thick, tough), the arachnoid membrane (weblike) and, attached to the brain and spinal cord, pia mater (tender matter). There are blood vessels running through the spaces between membranes, which are filled with CSF – actually, the ventricles of the brain drain into what's called the subarachnoid space between the arachnoid and pia. This three-layer, fluid-filled arrangement allows the CNS to float and protects it from shocks. If the meninges were not present, sudden shock caused by falls or blows to the head could generate forces within the brain, tearing it.

And you might want to know that meningitis is a dangerous bacterial infection in the meninges.

While talking about protecting the CNS, you should be aware of the blood-brain barrier and its parallel, the blood-spinal cord barrier. Brains need a good supply of blood to provide glucose, oxygen and other vital ingredients. Two problems: first, the composition of the blood varies during the day – glucose concentration goes up after eating and falls between meals; second, there are chemicals in blood that you don't want to get into your brain. The blood-brain barrier (BBB) isn't a bag surrounding the brain but a means for sheathing all of the blood vessels that penetrate every part of it. It's made of multiple cell types, making a barrier that allows small soluble molecules to simply diffuse across, but larger molecules are excluded unless they have a specialized transport system to cross the barrier. These larger molecules include glucose, which is picked up by astroglia that can feed glucose to neurons on demand, acting as reservoirs to buffer brain cells against the ups and downs of blood glucose concentration.

Before we start on the structure of the spinal cord, we need to briefly go through some common terminology relating to position. These are words you'll find a lot in nervous system anatomy, and their meanings are always the same.

Anterior and Posterior – front and back, ahead and behind.
Rostral and Caudal – also front and back: hint – the caudal fin is the one at the end of a fish, the one that propels it forward through the water.
Medial and Lateral – toward the middle, toward the edge.
Dorsal and Ventral – above and below/front and back. A dog's back, like any animal that walks on four legs, is dorsal; its belly is ventral. But humans walk upright, so the dorsal surface is where the spine is – back – and the ventral surface is the other side – front.
Superior and Inferior – above and below, not better or worse.

These terms are used a lot on their own and combined to give further positional information, as in rostromedial (front and central), anterolateral (front and wide) or posterolateral (back and wide). And two last things: first, afferent and efferent. These are terms used to indicate coming toward (afferent) and going away from (efferent – think of E for exit). Second, the word nucleus is context dependent. For us, it can mean the nucleus in a cell, or in anatomy, it can mean a group of cells organized together – we'll see many distinct brain nuclei as we go on.

The structure of the spinal cord, like that of the brain, is complex. Figure 1.3 illustrates its structure. It has four sections: cervical at the

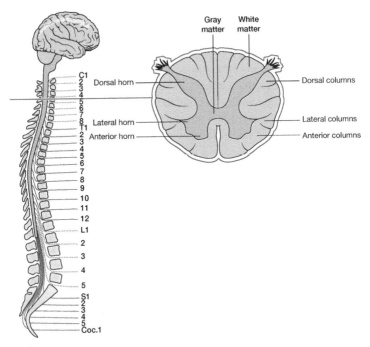

Figure 1.3 The spinal cord as it passes down the spinal column. The letter s and numbers are indices of position. The letter C stands for Cervical, T for Thoracic, L for Lumbar, S for Sacral and Coc for Coccygeal. The cross-section shows the relative positions of areas of spinal cord gray matter. The dorsal horn is also known as the posterior horn; likewise, the ventral horn is also known as the anterior horn. The Rexed laminae (not shown) run up and down the length of the spinal cord. In the cross-section here, Lamina 1 is at the tip of the dorsal horn, with the subsequent laminae running through to the bottom of the anterior horn.

top and then thoracic, lumbar and sacral, regions defined not just by location but by connectivity. They're structurally similar. If you were to cut across the spinal cord at any point, you'd see a butterfly-like pattern in the center. This is gray matter; what surrounds it is white matter. The difference is simple and applies throughout the CNS: gray matter is made of millions of cell bodies; white matter is myelinated fibers (myelin being fatty and white). Spinal cord gray matter has four columns of cells running up and down. These are the dorsal horn (aka posterior horn), the intermediate column, the lateral horn and the ventral horn (aka anterior horn). The dorsal horn receives somatosensory information – that is, information from across the body, like pressure, pain or warmth – which it transmits to the brain via white matter fibers. In contrast, the ventral horn sends output to skeletal muscles while the intermediate column and lateral horn send output to body organs.

The spinal cord gives us a first example of a more general problem. You'd think that the anatomy of the nervous system was pretty well fixed. In many ways, it is, but our understanding of the detailed cellular composition and connectivity of the brain is still developing, and changing theoretical approaches to function also create shifts in how we think about organizing and grouping structures. The Rexed laminae are a good example. Bror Rexed (1914–2002), a Swedish neuroscientist, saw how the traditionally defined nuclei in the spinal cord weren't the best way to understand it. He saw the spinal cord as grouped into laminae, areas defined by the organization of cells. Table 1.1 briefly describes the Rexed laminae: if you make a cut across the spinal cord to see the butterfly-shaped dorsal and ventral horns, lamina I is at the tip of the dorsal horn and lamina X in the ventral horn.

The pathways and tracts mentioned in Table 1.1 are made of white matter that surrounds the butterfly-shaped gray matter. They travel up and down the spinal cord and are grouped according to location: anterolateral, lateral and dorsal pathways, each containing fibers going between the brain and spinal cord, some in one direction, some the other and some in both. All of which very possibly sounds dull and not really relevant to biopsychology. You'd be wrong to think that.

We can start with an old favorite of textbooks and comedy, the reflex arc. In backboned animals, sensory information from the

Table 1.1 The Rexed laminae of the spinal cord

	Function	Location and connection
Lamina I	Pain and temperature sensing	Connects to brain via the spinothalamic tract
Lamina II	Pain perception	Connects to Laminae III and IV
Lamina III	Proprioception and light touch	Connects to Laminae IV, V and VI
Lamina IV	Non-noxious sensory information	Connects to Lamina II
Lamina V	Pain perception and other sensory information	Inputs to the brain via spinothalamic tracts; receives from the brain via corticospinal and rubrospinal tracts
Lamina VI	Spinal reflexes – it contains small interneurons	Sensory information from muscles; connects to the brain via spinocerebellar pathways
Lamina VII	Sends motor information to the body's organs	A large area that receives information from Laminae II–VI and body organs
Lamina VIII	Motor output to skeletal muscles	Varies along the length of the spinal cord; prominent in cervical and lumbar parts
Lamina IX	Motor output to skeletal muscles	Varies in size and shape along the length of the spinal cord
Lamina X		Surrounds the central canal; fibers cross over from right to left and *vice versa*

muscles and the internal organs of the body doesn't go direct to the brain but works by synapsing in the spinal cord. What this allows for is very fast reactions to stimuli. Sensory inputs arrive in the dorsal horns and can immediately activate motor outputs from the ventral horns. The brain will receive the same sensory information more slowly, but because an immediate reaction was required, the spinal reflex dealt with it: you step on a tintack, pull your foot away (spinal reflex) and then realize what's happened (brain activity). Reflex arcs are rarely monosynaptic – one sensory input, one motor output – and are usually polysynaptic, with interneurons working between input and output stages. In the context of biopsychology, this counts as important behavior but hardly seems to come within the orbit of 'mind'. While true, it masks an important point about nervous system organization that we'll return to, almost an Occam's Razor point: nervous systems need to keep everything simple in order not to waste energy on unnecessary activity and be able to make effective decisions quickly and accurately. Reflexes are the simplest form of do-it-now-reflect-on-it-later. Without such activity, behavior in the real world in real time would be seriously compromised.

THE STRUCTURE OF THE NERVOUS SYSTEM: THE PNS

The PNS includes all the nerves that run through your body. As we've said already, the nervous system is a complete package, so we need to maintain a balance between thinking in terms of the parts of a nervous system – brain, spinal cord, nerves – and an awareness of the unity of the whole. We'll see that in talking about the PNS because it joins neatly to both the spinal cord and brain. In this regard, there's another piece of anatomy to grasp: ganglia (plural: ganglion, singular). Ganglia are clusters of neurons interacting together and are found in all nervous systems, whether vertebrate or invertebrate. A brain is an organized collection of neurons segregated into anatomically identifiable structures with different parts having different functions and divided into two hemispheres, right and left. A ganglion has none of these features and is just a constellation of neurons working together. The ganglia we most often come across are spinal ganglia, which sit outside the spinal cord, not in it. The dorsal root ganglia hold the cell bodies of nerve fibers, bringing information in from the body (afferent neurons), and the ventral root ganglia have cell bodies that carry information out to the body (efferent neurons).

The PNS is typically divided into component parts that we'll go through one by one. First, the autonomic nervous system controls the internal body state. It's a system by which the CNS can control the body – cardiac muscle, the smooth muscle of internal organs, as well as glands – and sensory feedback to the CNS helps keep autonomic activity adjusted appropriately. The autonomic nervous system is divided into parasympathetic and sympathetic nervous systems. Each is a two-neuron chain, connecting through ganglia: there are preganglionic and postganglionic neurons. The parasympathetic nervous system has preganglionic neurons in the brainstem and lower spinal cord, which transmit to ganglia around the body at points close to their target organ. The postganglionic neurons go from the ganglia to that target organ and release the neurotransmitter acetylcholine there. (More on acetylcholine and the other neurotransmitters later.) The essential operations of the parasympathetic nervous system are described in two words: rest and digest. Meanwhile, the sympathetic nervous system has preganglionic neurons in the spinal cord, which project to ganglia immediately adjacent. The postganglionic neurons project from the ganglion to the target organ and release the neurotransmitter noradrenaline (aka norepinephrine). The two words for the sympathetic nervous system are fight or flight. As well as the autonomic nervous system, there's the enteric nervous system, which has complex ganglia in the walls of the stomach, small intestine, pancreas and gall bladder. It controls gut muscular activity, blood flow and secretions. Sympathetic and parasympathetic systems have a degree of external control over this, but inside the gut, the enteric system operates more or less independently. Autonomic and enteric nervous systems control and regulate the state of the body's organs. What about skeletal muscle, movement and the world outside the body?

It's the somatic nervous system that deals with these, using spinal nerves and cranial nerves attached directly to the brain, usually the brainstem. There are 31 pairs of spinal nerves, going right and left, leaving the spinal cord through openings in the vertebrae of the backbone. Each nerve subdivides such that the entire body is wired in. The spinal nerves are named after the parts of the spinal column where they're found. They are as follows:

- Cervical nerves: C1 to C8, going to the chest, head, neck, shoulders, arms and hands
- Thoracic nerves: T1 to T12, going to the back, abdominal and intercostal muscles
- Lumbar nerves: L1 to L5, going to the lower abdomen, thighs and legs
- Sacral nerves: S1 to S5, going to the legs, feet and genital areas
- Coccygeal nerve: Co1 – just the one – going to the tailbone

As well as the spinal nerves, there are 12 pairs (again, right and left) of cranial nerves directly attached to the brain. There is a mix of afferent information to the brain and efferent information from the brain. Some nerves only do one, afferent or efferent; others do both. While the spinal nerves are concerned with movement of the body, the cranial nerves have much more to do with the head and neck. Conventionally numbered using Roman numerals, the 12 cranial nerves are listed in Table 1.2.

Table 1.2 The cranial nerves

	Direction	Functions
I. Olfactory	Afferent	The sense of smell
II. Optic	Afferent	Visual input from the retina
III. Oculomotor	Efferent	Eye movements; pupil constriction, accommodation (focusing), eyelid opening
IV. Trochlear	Efferent	Also eye movements – there are different nerves for different eye muscles
V. Trigeminal	Afferent and efferent	Facial sensation; chewing
VI. Abducens	Efferent	Eye movements
VII. Facial	Afferent and efferent	Face movements, taste (from the front of the tongue), tears, salivation, eyelid closure, auditory reflex
VIII. Vestibulocochlear	Afferent	From the structures of the ear: hearing and balance
IX. Glossopharyngeal	Afferent and efferent	Taste (from the back of the tongue), blood pressure, elevation of pharynx/larynx (important for speech)
X. Vagus	Afferent and efferent	Vagus comes from the same Latin root as vagabond and means wandering: it's involved in taste, swallowing, soft palate elevation, speech, cough reflex, blood pressure and innervation of the lungs, heart and gut
XI. Accessory	Efferent	Swallowing, head and shoulder movements
XII. Hypoglossal	Efferent	Tongue movement

You might again be asking what all this has to do with biopsychology, and it's still a good question. We said in the introduction that psychology was concerned with both behavior and the mind. Behavior necessarily involves control of the body; mental life, while obviously depending on the brain, needs more than this. But recall Hippocrates and the four humors: the actual science we can set aside, but the fact is that how we think about, for example, emotions still involves an awful lot of body physiology. The cardiac palpitations of romantic love, the sweaty palms of anxiety: everyone experiences changes in their physiology that are part and parcel of emotional life. We know that all this peripheral anatomy can seem dull or confusing – or both – but it is important to have some idea of what's going on around your body and how it affects what you do and what you experience, emotionally and, as it's going to turn out, cognitively.

THE STRUCTURE OF THE NERVOUS SYSTEM: THE BRAIN

Patrick Winston (1943–2019), a computer scientist at the Massachusetts Institute of Technology, made an astute observation about the brain. "Everything is all mixed up, with information flowing bottom to top and top to bottom and sideways too. It is a strange architecture about which we are nearly clueless" [1]. That accurately describes the detailed wiring of the human brain – and indeed any other vertebrate brain. However, in anatomy, the power of your microscope is critical. We're going to pull up and focus at a different level and talk about the major brain structures.

In sections cut through a brain, the first thing you can see with the naked eye is white matter and gray matter. White, as we've said already, is made of myelinated fibers; gray is masses of cell bodies where there's no myelin – the insulation is only for axons. The ratio of gray to white is more-or-less invariant across vertebrate brains, and certainly true in the human brain: 50:50. Half the total volume of a brain is white matter. There are massive white matter fiber connections coming up the spinal cord into the brain, and within the brain, there are more big ones. Two of the most important are the corpus callosum, a sheet underneath the cortex that connects across the right and left hemispheres and the cerebral peduncles, which run from the frontal and temporal lobes of the cortex down toward the brainstem, cerebellum and spinal cord. It makes up half the volume of the brain, yet white matter has been almost wholly neglected in biopsychology, as if the wiring of the brain was of little consequence and all the important psychological business was only conducted in various groups of cell bodies – gray matter. But attitudes about the importance of white matter are changing. Genomic analyses have identified genes specific to white matter, which appear to be involved in intellectual dysfunction, depression, schizophrenia and autistic spectrum disorders. We can confidently predict that in 10 or 20 years, more will be known about the biopsychological functions of white matter, pointing to possible solutions to a range of mental health conditions.

You'll have noticed some alternative Latin terms for anatomical structures that we preferred to name in English, and there is a lot of overlap between Latin and Greek terminology and English. More problematic is the way in which terminology goes out of style. An older generation of neuroanatomists, classics scholars all of them, divided the brain in Greek into the rhombencephalon – which further split into metencephalon (the cerebellum and pons) and myelencephalon (the medulla oblongata) – mesencephalon (the midbrain), rhinencephalon (the olfactory structures) and the prosencephalon – which further split into the diencephalon (thalamus and hypothalamus) and telencephalon (the cortex, corpus callosum, basal ganglia and limbic system). Some of these are still in use, at least to an extent. Neuroscientists still refer to structures as mesencephalic, diencephalic or telencephlic, but the rest have largely gone. This surviving usage is why it's important to know about these older terms. However, as far as possible, we'll use English terminologies, or at least terms that never go out of fashion. So here are some of the major structures of the brain, back to front. We've tried to keep this simple, giving you basic foundations that you can build on as and when you need to. Figure 1.4 highlights the relative positions of some of the major structures.

The brainstem: divided into the medulla oblongata, where the spinal cord joins, and the pons, a bridge to the midbrain. The central gray (aka periaqueductal gray) runs through the center of the brainstem, an extension of the spinal cord. Along the length of the brainstem are nuclei where the cranial nerves enter and exit, several well-defined relay nuclei (for example, the parabrachial nuclei) and the reticular formation, a dense networks of neurons, separated into

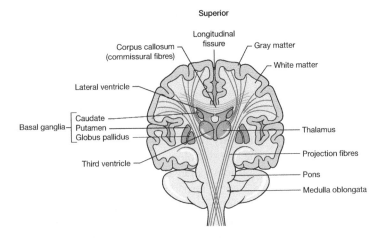

Figure 1.4 A cross-section of the brain showing the positions of some key subcortical structures – the basal ganglia and the thalamus.

the pontine reticular formation and the medullary reticular formation. These are not to be confused with the ascending reticular activating system, something dear to the hearts of many biopsychologists. It includes the locus coeruleus, the raphe nuclei and the pedunculopontine tegmental nucleus. It also includes elements of the midbrain and forebrain, which we'll talk about in a moment. All the nuclei of the ascending reticular activating system contain relatively small numbers of neurons. They all have incredibly long axons projecting down into the spinal cord but, more importantly, up into every part of the brain – this system literally has branches everywhere, regulating the activity of the entire brain, often through the release of monoamine neurotransmitters. Functionally, it's brought to you by the letter A: asleep, awake, alert, activated, aroused, attentive.

The cerebellum: this sits on top of the brainstem looking vaguely like a big cauliflower floret. It has three lobes: the anterior, posterior and flocculonodular. By volume, the cerebellum isn't the largest part of the brain, but if you count the number of neurons, the cerebellum is by far the biggest, home to 80% of all brain neurons in virtually all vertebrates. It's true that cerebellar neurons are small but nevertheless, this is an astonishing statistic. Its structure is complex.

There is an outer layer of gray matter called the cerebellar cortex, which has three layers: the molecular layer on top, then the Purkinje cell layer and, underneath, the granular layer. There is a highly structured neuronal organization across these cell layers, including inputs from a structure outside the cerebellum, the inferior olivary complex, in the brainstem. Within the core of the cerebellum, there is a lot of white matter and three pairs of deep cerebellar nuclei: the fastigial nucleus, the interposed nucleus (featuring the globose and emboliform nuclei) and the dentate nucleus. Inputs from outside the cerebellum arrive in the deep cerebellar nuclei and cerebellar cortex; output all goes through the deep cerebellar nuclei, as well as the closely related vestibular nuclei in the brainstem. Functionally, the cerebellum was thought to be concerned with motor coordination, and indeed it is. However, its function goes way beyond this into all sorts of sensory and cognitive operations. The complexity of its processing and the ubiquity of its functionality are increasingly well recognized and the subject of much investigation, theoretical and clinical.

The midbrain: conventionally divided into two areas: the tectum at the top and the tegmentum at the bottom. The tectum contains the superior colliculus and the inferior colliculus: the inferior one is slightly behind and below its partner. The superior colliculus has a lot of sensory input, but by far, the most dominant is vision, a direct connection from the retina. It also has direct output into motor control elements of the brainstem, making it a part of the brain that can initiate very rapid reactions to visual inputs. The information processing involved makes this more than just a reflex, but a similar rapid response capability is there. The inferior colliculus is in the same position with regard to hearing, though the auditory information it gets comes from nuclei in the brainstem (mainly the cochlear nucleus) rather than direct from the ear itself.

The tegmentum contains various nuclei. The two most closely studied are the ventral tegmental area and the substantia nigra, divisible into pars compacta (a cell-dense area) and the pars reticulata (less dense, more net-like – and do note, some people use zona instead of pars for these). The ventral tegmental area and pars compacta are among the most studied parts of the brain. They contain dopamine neurons that have been linked to reward and reinforcement (we'll talk about these later) and, clinically, have been shown to degenerate in Parkinson's disease, a debilitating motor disorder. The overactivity of these dopamine neurons has been linked with schizophrenia. You can see why these two small structures attract interest.

The thalamus: moving from the midbrain into the diencephalon, two important structures. First, the thalamus. The word comes from Greek and means an inner chamber, which is what this is. In the heart of the brain, it has sensory input from everything bar olfaction. It's divided into multiple discrete and identifiable nuclei, which have sensory, motor or integrative functions. The thalamus is typically regarded as the gateway to the cortex because it controls almost all sensory input going there.

The hypothalamus: another structure, another word. 'Hypo-' as a prefix in this context means below: the hypothalamus is below the thalamus. Like its overhead neighbor, the hypothalamus is divided into multiple nuclei with identifiable functions. Wired to multiple structures, from cortex to brainstem, as well as to the pituitary gland – there are direct connections through the pituitary stalk, by which the pituitary gland hangs below the brain – the hypothalamus is intimately concerned with the state of the body and the behaviors required to maintain it, like feeding, drinking and temperature control, as well as behavior required by the species, such as sexual activity.

The basal ganglia: it would be nice if there was some logic here, but there isn't. We talked about ganglia in relation to the spinal cord. The basal ganglia aren't like that at all. Instead, they're a collection of integrated structures: the caudate nucleus, the putamen, the globus pallidus, the subthalamic nucleus (below the thalamus, above the hypothalamus). The substantia nigra (both pars reticulata and pars compacta) and the pedunculopontine tegmental nucleus are often accommodated with the basal ganglia because of the closeness of their connections to the other elements of it. As you'll gather, the basal ganglia extend quite a way, from forebrain and diencephalon into midbrain and brainstem. So how come we think of it as one thing? First, you need to appreciate that the basal ganglia exist in more or less the same configuration in all vertebrate brains. Lampreys, the jawless fish and about the oldest vertebrate animals, have basal ganglia in the same arrangement as primates and all vertebrates in between. Basal ganglia evolution predates that of the cortex. So what are they for? Once thought of as only concerned with movement control, it's now clear that the basal ganglia have sensory, motor and cognitive functions. They help solve what is known as the 'action selection' problem. At any given time, an animal may have the choice of doing this, that or the other thing, or nothing at all. Choosing and doing is a simple way of thinking about basal ganglia function.

The basal forebrain: includes several modestly sized structures, such as the diagonal band of Broca, substantia innominata, the septal nuclei and the nucleus basalis of Meynert. There is yet another set of anatomical terms that overlap with those we've been dealing with. The basal forebrain often has the nucleus accumbens incorporated into it, but it's more common to find the nucleus accumbens described as the major part of the ventral striatum (the dorsal striatum being the caudate nucleus and putamen). The terms dorsal and ventral striatum came into use when neuroanatomists realized that there was a common pattern of connectivity in this part of the brain, a system of loops – the corticostriatal reentrant loops – that connect the frontal lobe of the cortex with the dorsal and ventral striatum, which then connect to the dorsal and ventral pallidum (that is, the globus pallidus and the tissue below) and these project into the thalamus. The thalamus projects back into the cortex. Through these structures, functionally segregated loops run in parallel (though they interact at key points) and operate as both continual feedback/feedforward and integration.

The amygdala: is a subcortical structure but sits embedded deep in the lower part of the temporal lobe. The name comes from Greek and means almond, the shape of which it's supposed to resemble. It has widespread connections to, among other places, the hypothalamus, ventral striatum, thalamus, olfactory system and brainstem. The amygdala has been associated with emotion for a long time – fear in particular – but is involved in more than this, with a role in learning and memory, for example.

The hippocampus: is among the most closely studied parts of the brain, and it has a complex structure that includes the dentate gyrus and areas CA1 and CA3 of the Cornu Ammonis, Ammon's horn. As with the amygdala, there is an imaginative stretch: hippocampus is the scientific name for a seahorse, which is what this part of the brain is supposed to look like. It sits like a saddle around the thalamus, separate both from it and from the cortex above, and is connected to both cortical and subcortical structures. It is always associated with memory. In 1953, Henry Molaison underwent brain surgery in an attempt to control epileptic seizures. Regrettably, when he recovered, he had utterly lost the ability to form new memories. To maintain his anonymity throughout his lifetime, he was known only as patient HM and was closely studied. The tissue destroyed by the surgery included much of the hippocampus and more besides. Somewhat simplistically though, at a time when brain centers for controlling specific processes was the theoretical standard, what stuck was 'hippocampus' and 'memory loss'. Different parts of the hippocampus are associated with different functions – stress and the ventral hippocampus, for example – but the memory element has evolved. What the hippocampus is best known for is spatial memory and its use, navigating animals through familiar and novel environments.

The allocortex: when we talk about the cortex, we're not talking about one thing. What people usually mean is the neocortex, all the material visible once the skull and meninges are out of the way. The allocortex is older and simpler than the neocortex. Structures identified as allocortical include, among others, the pyriform cortex, olfactory bulbs and elements of the limbic system, including the amygdala and hippocampus. What often also gets included in the allocortex is the juxtallocortex – principally the cingulate cortex. What's important is that the cortex isn't a uniform thing and that allocortical is used as a descriptor for a number of important structures that, as biopsychologists, you'll read about.

The limbic system: is a part of the brain well-known beyond science. The name comes from *la grande lobe limbique* – limbus is Latin for border – a description French neurologist Paul Broca (1824–1880) gave to a band of interconnected structures, including the cingulate cortex, hippocampus, amygdala, parts of the hypothalamus and some of the thalamus. Other anatomists later added more, such as the septal nuclei. Paul Broca did not speculate about its function. That waited for James Papez (1883–1958), who, in 1937, proposed that it controlled emotion. All this solidified: 'limbic' equals 'emotion' became an accepted equation. The idea that these structures form some unique bond isn't supported by contemporary neuroanatomy, and the idea that emotion is all regulated by these structures is implausible. Biopsychologists need to know about the limbic system, and biopsychologists should be wary of using concepts formulated a century and more ago to describe how we think about brains now.

The neocortex: organized into four major lobes – frontal, temporal, parietal and occipital lobes – the neocortex (aka, occasionally, isocortex or neopallium) is regarded as the pinnacle of brain development. The major lobes of the neocortex are shown in Figure 1.5.

It is folded into sulci and gyri (grooves and ridges, respectively; singular sulcus and gyrus – particularly deep sulci are called fissures). Folding has the effect of enabling a lot more tissue to be packed into a relatively confined space and, unsurprisingly, the human brain is the most heavily folded. The rat neocortex, for example, while sharing many of the cardinal features of a human neocortex, is totally smooth. The folding of the human brain creates an apparent depth of tissue, but the actual thickness of the cortical sheet of cells is between 1 and 4.5 mm, at an average of about 2.5 mm – and don't forget, a lot of white matter fiber contributes to the overall thickness of the cortex. The neocortex has six cell

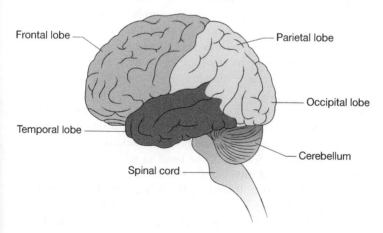

Figure 1.5 An external view of the human brain showing the major cortical lobes.

layers (with occasional exceptions: layer IV is largely absent from the motor cortex). From the top, these are as follows:

I. Molecular layer – relatively low cell density; cortical and subcortical inputs
II. Outer granular layer – projections to other areas of the neocortex
III. Outer pyramidal layer – projections to other areas of the neocortex
IV. Inner granular layer – receives thalamic input; short connections to other layers
V. Inner pyramidal layer – projections to subcortical structures and spinal cord
VI. Polymorphic layer – projections to subcortical structures and spinal cord

The neocortex also has cortical columns running across these six layers with a more-or-less common organization, often referred to as canonical microcircuits. (Canonical is an odd choice of word; it usually refers to the Canon law of the church but was adopted by mathematics where it means a general rule or formula: that's what's intended in the brain: ensembles of neurons that follow a general rule in their connectivity.) As we've already found, the cortex is connected by inputs and outputs to the subcortical structures below. Those inputs are not uniform and contribute to the functional separation of different parts of the neocortex. Broadly speaking, it's organized with motor functions at the front (the motor cortex is in the frontal lobe) and sensory functions at the back – vision in the occipital cortex, auditory perception in the temporal lobe. The cortex in between is often called the association cortex, where information coalesces together. A final neuroanatomical illustration, Figure 1.6, shows a section of the brain in which the relationships between cortical and subcortical structures are made plainer.

Two last things: first, Korbinian Brodmann (1868–1918) mapped the neocortex into 52 discrete areas, each one numbered. Brodmann areas are still in use, but it's important to keep in mind that Brodmann mapped on the basis of cell structure, not function. Second, it's easy to be seduced into believing that all-important human functions belong to the neocortex. While no one can doubt the importance

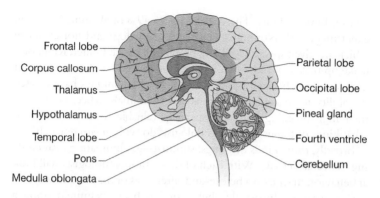

Figure 1.6 This illustrates the brain as if it had been cut in half along the long axis. It makes visible lobes of the cortex and key subcortical structures.

of the neocortex, it's important never to lose sight of the fact that everything it does involves cooperation with the rest of the brain and that assigning complex functions to small areas of the neocortex, while once all the rage, is no longer viable.

IN VIVO EXPERIMENTS ON ANIMALS

In vivo/in vitro/in silico: in life/in glass/in a computer. Experiments done on living animals are called *in vivo* experiments; those done in the proverbial test tube using cells or tissues taken from an animal are called *in vitro*, and computer modeling is referred to as *in silico*. *In vivo* experiments raise ethical issues that are beyond the scope of this book. All we will say is that in many countries, there is legislation regulating the use of living animals in research that will require scientists to justify what they do and submit their proposals to independent professional scrutiny. What drives *in vivo* biopsychological research in universities and hospitals is the need to understand how brain and body physiology shapes behavior and psychological processing – cognition, motivation, attention and so on. Such research can be informed by *in vitro* work and by computer modeling, but ultimately, if you need to find out about actual behavior, then experiments with living animals are needed.

In the United Kingdom, the four most commonly used animal species are mice – which are far and away the most widely used – followed by fish, birds and rats. These make up 96.49% of all animals used in experiments [2]. Nonhuman primates, cats, dogs and horses (all of which are given special legislative protection in the United Kingdom) make up 0.59% of those used. Mice have moved far ahead of the rest in recent years because molecular biological techniques have created the ability to make genetic alterations to animals. Mice, because of their high breeding rate, are especially valuable for such work. So what kinds of things do biopsychologists do with *in vivo* research?

Broadly, three kinds of things: stimulating, eliminating or recording neuronal activity. With each of these, biopsychologists will look at behavior, maybe on a before-and-after basis or maybe while events are occurring. Historically, behavior has been examined using a variety of different kinds of mazes and by operant work – pressing levers for a specific outcome in an operant box (aka a Skinner box, after B. F. Skinner, who developed it). Increasingly, virtual environments are being used, in part at least because of a growing belief that artificial environments like mazes and Skinner boxes do not properly capture natural behavior. Virtual reality experiments – displays with touch screens – are thought to mimic real life more closely.

Stimulating neurons can be done by drug administration, given through the body by injections, or by directly introducing chemicals into the brain via surgically prepared cannulae. Eliminating neuronal activity was, until recently, typically done by surgically destroying a specific part of the brain or by temporarily inactivating it with a local anesthetic. However, contemporary research increasingly uses animals that have been genetically modified in very specific ways. This might involve genetically modifying animals so that particular features of particular neurons never develop and specific neurons are 'knocked out'. Or a technique called optogenetics may be used. This involves using molecular biological techniques to insert light-sensitive molecules into targeted neurons. When light is shone onto these neurons using surgically implanted probes, neurons will either become active or inactive, developing on the genetic modification that has taken place. It offers the ability to target neurochemically identifiable neurons within a mixed population, enabling the relationships between behavior and very specific groups of neurons to be examined.

Recording from neurons has changed dramatically in the last few years. Historically, recordings from the brain using electrodes would involve only a tiny number of neurons. A technological jump came with the development of multitrodes, electrodes with a series of independent sites along their length, enabling multiple recordings to be made. A new technology, Neuropixels, has an array of electrodes that enables over 5,000 neurons to be recorded simultaneously from a wider neural space than was imaginable only a short time ago. In parallel with this astonishing increase in recording capacity, there have been advances in the ability to decode patterns of neuronal activity. This is genuine frontline material. In the past, when scientists made recordings in sensory systems, they would find activity clearly associated with a particular sort of stimulus. What they also got was a lot of extraneous activity coming from the neurons, which they thought was random noise. Better recordings and better analyses now confirm that this wasn't just noise but neuronal activity driven by the behavior of the animal under test. That is, both sensation and current behavior are driving the electrical activity of neurons. This result wasn't expected and is game-changing for how we think of information processing in the brain, a serious theoretical advance driven by *in vivo* research with mice.

HUMAN EXPERIMENTS

Invasive experiments involving the human brain are rare and, like *in vivo* experiments, are subject to independent ethical review. Very occasionally, a patient undergoing therapeutic neurosurgery will have recordings taken from electrodes, but it's far from common and, in truth, contributes little to our understanding of biopsychology. Noninvasive experiments, on the other hand, can be revelatory. Electroencephalography, EEG, involves placing electrodes on the scalp and recording widespread electrical activity from the brain below. Electroencephalography is the process of doing this, and an electroencephalogram is the recording made. In terms of assessing psychological processes, EEG has deficiencies and virtues. It lacks precise anatomical specificity, and it is difficult to reach the deepest parts of the brain. The virtue is speed; recordings of brain activity can be made in real time. In contrast, the reverse is true for neuroimaging: precise localization, lower realization in time.

Neuroimaging comes in many forms. There is the simple taking of an image of brain structure using computerized axial tomography

(CAT; scanning, usually now abbreviated to CT scanning) or magnetic resonance imaging (MRI). Positron emission tomography – PET scanning – goes a little further by using harmless radioactive tracers injected into the bloodstream and which are detectable in the brain. Using different tracers enables clinicians and scientists to examine different sorts of events, like metabolic rate, the density of specific types of neurotransmitter receptors and blood flow. Again, more specific than CT or MRI, diffusion tensor imaging allows for what's known as tractography, mapping the white matter fiber systems in the brain in detail.

As well as mapping the physiological structure of the brain, there are methods for stimulating it. Transcranial magnetic stimulation uses magnetic fields to activate neurons in experimental studies and in the clinic, where it has been used in the treatment of depression. Similarly, transcranial direct current stimulation applies current via electrodes positioned on the head to excite or inhibit neurons in the cortex. Deep brain stimulation is different because the electrodes used are actually implanted in the brain rather than only being on the scalp. Deep brain stimulation began as a treatment for intractable drug-resistant Parkinsonism – a debilitating neurodegenerative movement disorder – but has been applied in other clinical contexts.

The best-known and most widely used method for exploring the function of the human brain is functional magnetic resonance imagining (fMRI). MRI scanning looks at brain structure; fMRI looks at what it's doing, using the blood-oxygen-level-dependent signal (the BOLD signal) as an index: more signal, more activity in that part of the brain. People placed in scanners can take psychological tests; the images generated by the BOLD signals are thought to show where that psychological function is being processed. fMRI is a widely used technique to look at brain activity, but it is not without problems. Questions have been raised about reproducibility and replicability: can different analysts get the same result from one set of data (reproducibility), and do different studies generate mutually supportive conclusions (replicability)? The predictive clinical usefulness of fMRI is under a shadow, and there have been questions raised about whether or not groups of people – differentiated by sex, gender, socioeconomic status, education or ethnicity, for example – consistently give the same sorts of results. A very new technology, resting state fMRI doesn't look for brain activity prompted by a psychological task but at the brain's resting state activity, with a view to understanding how brains are organized into networked systems. It's promising, but it seems that, like fMRI, there is greater individual variability than might have been expected.

THEORY, PRACTICE AND EXPECTATIONS – CHALLENGES

What we have tried to do here is provide a framework for understanding the structure of the nervous system, which is obviously complicated, sufficiently so for a lifetime of investigation. The structures of the brain, its geography if you like, are clear and we've outlined the major ones in this chapter. The composition of the brain, on the other hand, remains a subject of discovery, with new features of neurons and glia constantly being described. Molecular biology and the revolution in genomics are a continual source of new information. Moreover, the composition of the brain remains open to new interpretations in other ways. How structures are grouped and organized, has changed as more about the nervous system is discovered. Recall Bror Rexed's changed approach to spinal anatomy or the way in which new discoveries about corticostriatal reentrant loops changed the way in which we see the cortex connected to structures below. Neuroscience evolves: so what does this say about core business for biopsychologists?

Francis Gall's phrenology gifted a legacy of centers – the idea that different psychological functions would be located in discrete parts of the brain. Attempts to identify brain centers persisted for a long time. Scientists tend now to think in terms of networks rather than centers, but there is still a continuing sense of specific brain functions in specific places. There are some big philosophical questions embedded in this. We can start with these:

(1) Does the way in which we describe our cognitive and emotional lives, either in terms of everyday language or the more rarefied talk of academic psychology, actually map directly onto the brain? Does how we describe our psychological experiences mean that brain function exactly matches it, or are brain information processing and our personal experience two different things?

(2) What does an account of (for example) human memory or perception in terms of neuron firing actually mean? Are we saying that when these neurons in this place are active, we experience something? Or is there what philosophers have called an explanatory gap between physical activity in the nervous system and our experience of what something is like?
(3) Does the way in which we describe our cognitive and emotional life transfer to animals? Do all species have the same sorts of processes happening? Is there anything about cognition and emotion that is uniquely human, something that other animals don't have?
(4) Are cognition and emotion all locked into the brain and totally accounted for by the activity of its neurons, or do they extend into the body and even out into the world? Does culture have a role to play?

We'll explore these more as we go through the book. For now, we need to move on to look at other parts of the body and how they might make contributions to biopsychology.

CHAPTER 1: SUMMARY

- *The brain is of special interest to biopsychologists, but it's important to understand that the nervous system of a body is a single complete package made up of the CNS (brain and spinal cord), the PNS (including the autonomic nervous system [divided into sympathetic and parasympathetic nervous systems], somatic and enteric nervous systems) and the cranial and spinal nerves.*
- *The essential cells of the nervous system are neurons, supported by glial cells; nerves are made of multiple neurons bound together. The brain is divisible into right and left hemispheres, gray and white matter and is described in varied terms, some relatively antique (derived from Latin and Greek) and some more modern. In detail, brain anatomy is exceptionally complex.*
- *Brainstem, cerebellum, midbrain, thalamus, hypothalamus, basal ganglia, basal forebrain, amygdala, hippocampus, limbic system, allocortex, neocortex – if you understand where these are and how they interact with each other, you will have a good basic grip of the structure of any vertebrate brain.*

- *Knowledge about the brain is gained from controlled and regulated experiments involving living animals as well as studies of human brains, often using sophisticated neuroimaging techniques.*
- *There are important philosophical questions about the relationships between psychological states and physical states in the brain. Do physical events (such as the activity of neurons) fully explain psychological states, or is there what's called an explanatory gap?*

2

PHYSIOLOGY FOR PSYCHOLOGISTS

What happens inside the body is important for psychology in both directions: how psychological states affect physiological processes and how physiological states affect psychology. In this chapter, we will run through important physiological systems and how they contribute to biopsychology. What we'll be looking at are interoceptive as opposed to exteroceptive systems – that is, signaling that arises from inside the body rather than that which comes from outside. Vision, hearing and the rest of the exteroceptive systems we'll deal with later on when we consider cognition.

In the last chapter, we looked at the structure of the peripheral nervous system (PNS). What needs to be emphasized is that this system is not passive, and it doesn't only send untreated interoceptive or exteroceptive information into the brain. Preliminary analysis and selection, collation, and sorting of information happen *en route* to the brain in ganglia and in sensory organs. The retina, for example, does not simply send a pattern of pixels to the brain. Instead, there's first-pass analysis of things like movement and light intensity so that what goes along the optic nerve to the brain is partially analyzed data rather than photographic images. Likewise, information arriving from the body's organs via the vagus nerve is precoded in the ganglia of the PNS. What arrives in the brainstem from the vagus is not raw data but has already been sorted and coded to some extent. And of course, we know that spinal reflexes can generate behavior without reference to the brain, just-do-it responses when speed is of the essence. It's important to understand that the PNS is not a neutral delivery service but an active

DOI: 10.4324/9781003215509-3

participant in biopsychological processing. When we think about brains and bodies, we must consider that what happens inside the body is more than a contribution to what's going to happen in the brain but something that is constitutive, an integral part of psychological processing. With all that said, we can look at different body systems that are important for biopsychology. They are not wholly independent but act together to maintain the body while still having independent lives of their own.

First, **the cardiovascular system**: the heart, blood vessels and blood itself. The heart has four chambers: two atria (singular, atrium) and two ventricles. These are illustrated in Figure 2.1. Valves between them make the blood go only in one direction: the right ventricle pumps blood through the pulmonary arteries (arteries go away from the heart) to the lungs, where small blood capillaries collect oxygen and eliminate carbon dioxide. Pulmonary veins (veins that come back to the heart) bring oxygen-rich blood to the left atrium. From here, blood goes to the left ventricle, which pumps

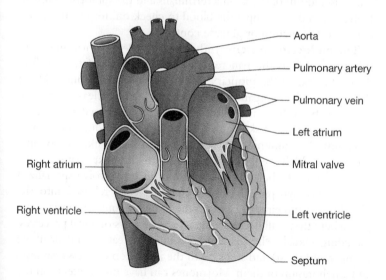

Figure 2.1 The human heart. As we make clear, what happens in the body has a significant influence on biopsychological activity. Changes in heart rate are associated with a variety of cognitive and emotional states.

it via the aorta to the rest of the body. Blood, now much depleted of oxygen, returns to the right atrium and from there is pumped to the right ventricle, ready to go on another circuit. This makes four chambers, two circuits: the pulmonary circuit collecting oxygen from the lungs, the systemic circuit delivering blood to the rest of the body. Blood gets to the brain via two pairs of arteries – the vertebral and internal carotid arteries. These divide to create more arteries that branch repeatedly, becoming small capillaries, delivering blood to every part of the brain because every cell in the brain needs what blood delivers: oxygen, glucose and more. Veins take blood away from the brain, merging to create larger ones called sinuses that run through the meninges. These sinuses drain into the jugular vein in the neck. Do remember what we talked about before: the importance of the blood-brain barrier in protecting neural tissue from unwanted material and daily fluctuations in blood composition. There are a few brain areas called the circumventricular organs where the blood-brain barrier is absent: area postrema, the organum vasculosum of the lamina terminalis and the subfornical organ. Here, neurons can sample the blood to check on, for example, evidence of dehydration or glucose content.

The endocrine system synthesizes and releases hormones from structures such as the pancreas, kidneys, heart and fat, as well as glands (including the pituitary gland, attached to the underside of the brain) and sex organs. The endocrine glands release chemicals inside the body. In contrast, exocrine glands release material outside it – sweat, tears and skin oils, for example. Hormones are a form of chemical communication between cells, but unlike neurotransmission – which we'll talk about more in the next chapter – the messages tend to be long distance and do not involve specialized connections: synapses. Hormones are typically released into the bloodstream and circulate around the body, finding appropriate targets where they will have specific effects on physiological processes, including sexual behavior and reproduction, eating and drinking (hunger and thirst) and stress, both the induction of stress responses and amelioration of them. Hormones can have direct effects on the brain and can be regulated by the brain, prompting release or damping down their actions.

The immune system is an internal protection system, illustrated in Figure 2.2. To defend the body, it deploys leucocytes

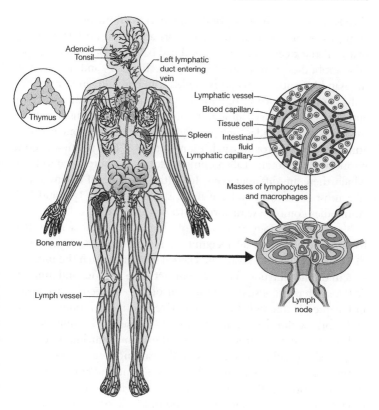

Figure 2.2 The human immune system, which is being increasingly recognized as important in biopsychology. The illustration shows how the presence of the immune system in the brain, where the glymphatic system, which drains into the lymphatic system of the body, has a key role in maintaining brain health.

(aka white blood cells); antimicrobial peptides (aka host defense peptides); cytokines, such as the interleukins; and lymphocytes. It uses the lymphatic system, thin tubes throughout the body where there are lymph nodes. The response to a wound of some sort comes by inflammation and the release of protective chemicals like histamine. The response may be localized, like a wasp sting, or more widespread, like a fever. Inflammation is relatively nonspecific, but the immune response is targeted. Lymphocytes come in two basic types:

T-cells (from the thymus gland) and B-cells (from bone marrow). Antigens (short for antibody generator) are found on lots of things that can attack cells, like viruses and bad bacteria. B-cells create specific antibodies in response to antigens. Both B- and T-cells have thousands of antigen receptors available, but an antigen will activate only a small proportion of lymphocytes – those with the matching antigen. The primary immune response comes after this matching: the lymphocyte divides, one part fighting the invasion, one part creating a fresh receptor to catch more antigens. The fighting is done by antibodies, proteins called immunoglobulins, abbreviated to Ig, of which there are different types IgM (the first line of defense), IgG (the most numerous) and the more specialized IgA, IgD and IgE. Antibodies combine with antigens to destroy them.

Neuroimmunology is an area of clinical and basic science that has expanded out of all recognition in recent years. It was once thought that the brain was a privileged site into which the immune system did not intrude. That is incorrect. The nervous and immune systems interact constantly with each other – indeed, as we noted in Chapter 1, microglial cells are derived from the immune system and work within the brain to protect and serve. Immune system cells such as cytokines interact with the brain, signaling states that need dealing with; stress, for example, can increase cytokine levels in the brain. Inflammation in the brain – neuroinflammation – has been associated with a variety of disorders, including Alzheimer-type dementia, Parkinsonism and multiple sclerosis. It's not all bad though: cytokine activity in the brain has been associated with the regulation of everyday psychological processing, body and brain acting harmoniously. A really significant recent discovery has been the glymphatic system, a waste elimination system unique to the brain that uses a system of channels around blood vessels to clear unwanted proteins and metabolites (the by-products of cellular metabolism). These drain out of the brain into the body's lymphatic system. The glymphatic system is particularly active during sleep, and there is a lot of current research examining its potential protective role in various neurodegenerative disorders.

In one sense, the **digestive system** is just a tube, mouth to anus, but that hardly captures the complexity of the organs involved. We don't talk about a tube but about the alimentary canal, which has distinct sections. First, the mouth and pharynx, where saliva starts to digest carbohydrates and a food bolus – a bolus is a small rounded mass – is made so that it can be swallowed. Taste processing (aka gustation) happens using sensory receptors on the tongue. This is important: tasty nutritious food initiates cephalic reflexes of digestion like salivation and insulin secretion that are preparatory for digestion. In contrast, cephalic rejection reflexes – spitting out food – are important for avoiding the consumption of bad food and drink. After this is the esophagus, which leads to the stomach. Food is moved along by peristalsis, a continual rhythmic flexing of the muscle wall of the esophagus. The enzymes in saliva continue to work on the bolus as it moves toward and into the stomach. This holds on to what's been eaten, releasing material bit by bit for processing, helping manage the constant demand bodies have for energy. Enzymes start to digest protein in a gastric environment (gastric means relating to the stomach) that is highly acidic (pH 2). A layer of mucus and a process of continual regeneration of the stomach wall protect the organ itself from this acidity. (As an aside here, we talk about 'the stomach' because that's what we have – one stomach – but you should be aware that other animals have different arrangements. Ruminants like cattle have three stomachs to deal with the amount of plant material they get through, while birds have a crop in front of the stomach and a gizzard behind to help them process and digest materials.) The stomach mixes food and releases what's called chyme into the small intestine. In humans, this is a 6-meter tract: the first 25 cm is the duodenum (into which the pancreas and gall bladder deliver chemicals to promote digestion), after which there's the jejunum and ileum. Through these, ingested matter is progressively broken down: proteins, fats and carbohydrates are degraded into smaller parts for use by the body. Blood capillaries and veins, as well as the lymphatic system, take nutrients out of the small intestine (sometimes by passive diffusion, sometimes by specialized molecular pumps), draining into the hepatic portal vessel, a major blood vessel that goes directly to the liver. After the ileum of the small intestine comes the large intestine (aka the colon). The caecum is also wired in at this point, making a three-way connection. The main part of the caecum is the appendix – the caecum *en bloc* is involved in cellulose digestion and is small in humans, who can have the appendix removed with no real consequence. The large intestine moves material quite slowly, extracting the last amounts of water from it before waste is expelled at journey's end, the rectum. Sphincters control expulsion: the rectal sphincter is involuntary, the anal sphincter not so. While in transit through the alimentary canal, food is broken down into usable parts, but this is just a part of the story. How is value extracted from food? What about water? And what about bacteria? Let's start with them and what's called the microbiome.

The microbiome is an expression used to describe all the bacteria in your body. Not so many years ago, it would be taken for granted that bacteria were simply bad and needed to be destroyed or at least controlled. Clearly, some bacteria do need to be avoided like the plague, metaphorically and literally: plague is caused by the bacterium *Yersinia pestis*. However, not all bacteria are bad, and many are beneficial. The human body contains vast numbers, to the amazing extent that there is actually more bacterial DNA in a person than there is human DNA. They are present everywhere, even in the brain, but those in the gut are of most interest here. They aid in the digestion of ingested material, maintain a healthy gut environment and signal to the brain. The vagus nerve – which has branches deeply embedded in the gut – picks up signals from bacteria and transmits them to the brain. Additionally, the brain is able to detect bacterial fragments carried in the bloodstream. Gut bacteria are known to be involved in metabolic syndrome, which combines obesity, diabetes and high blood pressure (hypertension), and in dysbiosis, which involves fatigue, inflammation and various gastric and urinary problems. The true extent of microbial involvement in human physiology is still being worked out, but it is already clear that it is more significant than previously credited.

Thinking about how value is extracted from digested food brings us to the liver. This is the largest internal organ of the human body and has a critical role in the storage and use of energy. The hepatic portal vein and the hepatic artery take blood into and out of the liver, which does several jobs. For example, it detoxifies blood and is a key storage site. Specific vitamins are stored – it has sufficient B12 to last a year – as well as glycogen. Glycogen, found in muscles as well as in the liver, can rapidly be activated by glucagon (from the pancreas) to provide energy for swift action – fight-or-flight reactions, for instance. Energy stored as fat can't be accessed this quickly. What's more, there are detectors in the liver for water and glucose that signal to the brain via the vagus nerve. This sensory information carried by the vagus works on the brainstem and hypothalamus, which, in turn, can respond to changes in liver activity. It would be hard to underestimate the role of the liver in controlling energy balance.

The liver is able to deliver energy for use by the body very quickly. How, on the other hand, is long-term energy storage and use managed? It brings us to fat. More technically, we talk about adipose tissue, which is found under the skin (subcutaneous fat), around internal organs (visceral fat), where it provides cushioning, and – perhaps most surprisingly – fat is found in bone marrow. Adipose tissue is contentious for three reasons: first, dietary fats. What we eat contains saturated fats, transfats, monounsaturated fats and polyunsaturated fats. The first two tend to be solid (like butter), while the second two tend to be liquid (like oils). In general, solid fats are considered worse for you and liquid ones better, but opinions are not as consistent as all that. (Some years ago, the use of butter declined under pressure from artificial products, but if you look at any celebrity chef cookery program now, you'll see a lot of butter being used.) What all fats deliver to the body are lipids that fat cells – aka adipose cells – store (as triglycerides) and release as required to fuel body activity. Fats are energy dense with nine calories per gram, compared to carbohydrates and proteins, which only have four. However, it's not only dietary fats that can be converted to body fat. Carbohydrate intake provides glucose that can be used immediately by cells, or converted to glycogen in the liver as a short-term store or converted to fat and stored longer term. The second point of contention is body image. Say 'fat' in conversation, and most people will assume you're not talking about physiology but about body shape and size. Susie Orbach captured this in the late 1970s with her book Fat Is a Feminist Issue. Body image and body shaming remain deeply contentious issues among adults and children. Fat is a loaded word that we can't avoid in talking about body physiology, but it's one that needs to be treated with appropriate respect. The third point of contention is the relationship between the white fat stored in the body as adipose tissue, brain activity and the regulation of metabolism. Previously it was thought that adipose tissue contained no nerves and no means of signaling to the brain. Neither is true. Adipose tissue is, in effect, an endocrine organ with innervation by nerves of the PNS that can both detect and influence activity. Adipose tissue also releases adipokines, which are types of cytokines like those released by other parts of the endocrine system. There are many adipokines, but the best studied is leptin, which acts on the brain and the gut, where it works synergistically with ghrelin, a hormone that stimulates feeding. Obesity is associated with an inability of the brain to detect leptin.

It's worth noting that what we've been discussing is white adipose tissue. There is also brown adipose tissue, brown because it contains more blood capillaries and more mitochondria (the energy generating organelles in cells). Unlike white adipose cells, which have a single lipid droplet, brown ones have multiple small ones. Brown adipose tissue – colloquially, brown fat – is found mixed in with white (and so sometimes called beige fat or brite fat, a mix of brown and white) but also independently. Its function is thermogenesis, the production of heat. Its presence in newborn human infants was long known, but it was assumed to disappear with age. Not so: it might diminish but is present in adults, where it has a role in mitigating overweight by burning excess energy intake as heat. Brown adipose thermogenesis isn't the only mechanism for controlling body temperature. The hypothalamus continually checks body temperature, and if there's significant variance from the normal 37°C (98.6°F), changes are made by altering muscular and metabolic activity. If body temperature is too high, sweating cools the skin.

Moving on from food and energy, what about that most essential substance, water? Controlling water balance is technically called osmoregulation and is complicated. Your body has four fluid compartments: inside cells (intracellular fluids – about two-thirds the total) and extracellular fluids, outside cells and further divided into three: cerebrospinal fluid (CSF, which we came across in the previous chapter and which doesn't contribute to water balance), blood plasma (the water content of blood, with various additions such as ions and proteins, but not blood cells: plasma is about 10% of the total body water content) and, finally, the fluid between cells – interstitial fluid – which is about a quarter of body fluid. Under normal circumstances these four compartments are kept apart by cell membranes that are semi-permeable (permeable to certain things under certain conditions). Because intra- and extracellular fluids are isotonic (they have the same contents), there's no pressure (osmotic pressure) on cell membranes to create movement across them. However, if extracellular fluids become hypertonic (losing water, becoming more concentrated), fluid is drawn out of cells, and vice versa: if extracellular fluids are too dilute (hypotonic), water is drawn into cells. Changes in the contents of fluid compartments are signaled to the brain, which will initiate physiological changes to protect water balance, as well as behavioral changes to get more water on board. Physiologists talk about osmometric thirst, which comes after cellular dehydration, and volumetric thirst, which is driven by intravascular dehydration. Both of these are complex processes.

It is the job of the kidneys to control osmometric thirst. They control the amount of water and sodium present in body fluids (sodium being a key part of sodium chloride, the most essential salt). Nephrons in the kidneys are connected by the ureter to the bladder, from where excess fluid is released as urine. The hormone vasopressin (aka antidiuretic hormone) acts on the kidneys to control how much water will be released. It's synthesized in the hypothalamus of the brain and transported to the pituitary gland from where it is released into the bloodstream. Vasopressin release from the pituitary is decreased by drinking enough water; dehydration increases its release. Sodium levels – important for physiology – are regulated in a parallel manner by aldosterone, released from the adrenal glands: too much salt, aldosterone levels fall; too little salt, aldosterone levels increase. Volumetric thirst is necessary to maintain blood. If blood volume is too high, it causes increased pressure in blood vessels (hypertension), which is a condition associated with, for example, stroke. Medics call hypertension 'the silent killer' because its presence is often not detected until something bad happens. If blood volume is too low (hypovolaemia), it can lead to heart failure. The kidneys detect loss of blood and secrete renin into the bloodstream. This is an enzyme that converts angiotensinogen (ever present in blood) to angiotensin I, which is converted to angiotensin II by an angiotensin-converting enzyme (ACE; ACE inhibitors are sometimes used to treat hypertension). Angiotensin II acts on the brain to stimulate vasopressin activity, which works to protect water content and prevent further loss. Angiotensin II also acts on the adrenal gland for the release of aldosterone, which contracts blood vessels to increase blood pressure – and it's a powerful stimulant for drinking. The heart is also involved in volumetric thirst in two ways. Baroreceptors detect loss of blood volume when blood returns after circulation; atrial natriuretic peptide is released if there is too much water, inhibiting renin, aldosterone and vasopressin, and suppressing drinking.

The physiology of water balance is necessarily complex, given how critically dependent we, and all other animals, are on it. Brains have a central role to play in osmometric and volumetric thirst. So, for example, atrial baroreceptors signal to the brain via the vagus nerve; angiotensin II has direct actions on the brain, for example, in the circumventricular organs where the blood-brain barrier is absent, and there are neurons in the brain called osmoreceptor neurons whose activity is changed by hydration. Networks of neurons, especially but not exclusively in the brainstem and hypothalamus, monitor water balance, trigger physiological changes and push behavioral selection mechanisms toward drinking.

Everything we've talked about physiology so far has been to do with maintaining and protecting the individual. Managing water balance and energy supplies, responding to wounds and harmful events. Everything has been about individuals, whether persons or animals, and how they stay fit and healthy. Sexual behavior is a little different. Going without food and water, failing to deal with injury or infection – individuals must avoid these or suffer the consequences. Going without sexual activity might have consequences but is not actually fatal. Sexual activity is practiced by individuals and is essential for survival of the species. Sexual activity is another contentious topic, and we are not going to engage here with issues around biological sex and gender. In strictly biological terms, the point of sexual activity is a means of reproduction that involves combining the genes of a male and female to create unique new individuals. It means that males and females, of whatever animal type, have different sex organs that enable the insemination of eggs by sperm, the internal development of embryos and fetuses and, after birth, the suckling of infants by lactation. (Other animals, of course, lay and incubate eggs externally and do not suckle their young.) Hormones govern a great deal of this. Reproductive hormones (aka sex hormones, sex steroids, gonadocorticoids or gonadal steroids) act at receptors on target organs. The sex hormones include estrogen, progesterone and testosterone, which are synthesized in the ovaries and testes. Other hormones, such as luteinizing hormone, which controls lactation, are synthesized in the pituitary gland. Reproductive hormones are essential, obviously, for reproduction, development, growth and puberty. Less obviously, they have a role in such things as white adipose deposition and inflammatory responses. They aren't constant: female sex hormone activity fluctuates across the menstrual cycle and changes during menopause. Stress can affect the activity of both male and female sex hormones.

This might seem like overkill, too much information about bodies with limited relevance to biopsychology. But think back to the opening chapter and how the ancient Greek physician Hippocrates believed that four body humors controlled psychological states. He wasn't right in specifics but the idea that what goes on in the body helps shape psychological life and behavior remains very important. Understanding our bodies is crucial for biopsychology. What are called primary motivated behaviors – feeding, drinking, temperature regulation, reproductive behavior – all rely on interactions between brain and body, with signals between the two using the PNS and a very wide variety of chemical substances released by body organs and carried through the bloodstream. Chemical signals include those taken on board from outside, such as glucose, water and oxygen, and those manufactured inside such as insulin, leptin, vasopressin, angiotensin II, reproductive hormones and cytokines. The interplay between the PNS and body organs involves the brain in two ways. It has to try and manage body physiology *per se* – for example, by secreting vasopressin to preserve water in the body – and at the same time initiate appropriate actions that will achieve a number of ends. In the context of feeding and drinking, for example, brains are involved in remembering in what places food is usually found, discriminating between good food and bad (animals quickly learn to avoid noxious substances through learned taste aversion), paying attention while capturing food (for carnivores in particular) and possibly returning to their homes with it, simultaneously scanning for threats. As well as all this of course there is eating, the physical act of consumption. To top all this off, brains have to engage mechanisms of action selection. When there are choices to be made to act or not to act, how do animals do it? The basal ganglia, as we saw in Chapter 1, are critical for this most basic of processes.

All of this requires effective use of both interoceptive and exteroceptive sensory information coming from inside and outside the body. It also requires cognitive processes such as learning and memory. Many scientists argue that these necessarily involve a fundamental integration with action – what's known as enactive cognition – drawing ever-closer relationships between what we think of as psychological processes and behavior. And we know that emotions involve physiological events: heart rate, sweating, control of bowel and bladder are all associated with emotional states. We need also to reflect on the fact that using our bodies is good for us. Aerobic exercise has beneficial effects on psychological processing, especially with age, when it might help stave off dementia, and in other clinical conditions such as depression or rehabilitation after traumatic stress.

In later chapters, we'll discuss topics including sensation, cognition and emotion in more detail. The important message from this chapter is that a comprehensive understanding of biopsychology necessarily involves appreciation of body physiology and the dynamic interaction between the central nervous system, PNS and all the body's organs. Right now, however, we need to move into a different but closely related space: psychopharmacology.

CHAPTER 2: SUMMARY

- *Sensation can be divided into exteroception and interoception, the detection of events outside and inside the body. Understanding the world is important in psychology to the extent that what happens internally can be easily neglected.*
- *Events in the PNS are important. What happens here is not simply a contribution to information processing but constitutive; that is to say, it's an essential component rather than an additional "helps to have".*
- *This chapter examined the structure and function of physiological systems: the cardiovascular system, the endocrine system (including reproduction), the immune system and the digestive system (including the management of water balance).*

- *Understanding the relationships between these systems and the brain is vital: the brain can regulate activity in all these body systems, and all these body systems have effects on the brain. Principally, it is impossible to understand how primary motivated behaviors work without this interactive understanding: feeding, drinking, keeping warm and reproduction depend on it.*
- *As well as critical for these basic behaviors, cognition and emotion are susceptible to changes in body state. Indeed, cognitive processes like perception, attention, learning and memory and decision-making are all, at root, designed to help animals survive and thrive by making sure, for example, that animals understand what food is, where it is, how to get it and, when necessary, how to defend and store it.*

THE CHEMISTRY OF THE BRAIN

PSYCHOPHARMACOLOGY

Psychopharmacology looks at the effects of both therapeutic and recreational drugs on behavior and psychology. The use of drugs to treat psychological conditions has been widespread but not without controversy. Do drugs treat and cure mental illness, or are they just palliative, relieving the symptoms without dealing with the underlying cause? The use of recreational drugs, persistent throughout human history, is an extraordinarily complex issue in medicine, health, social policy and law. It's universally agreed that drug addiction – indeed, any form of addiction – is a bad thing, but what is it, and how can it be effectively dealt with? In both therapeutic and recreational cases, it's important to understand that the actions of drugs are complex, with many psychological, behavioral and physiological factors changing their effects.

To start, what is a drug? It's a chemical, obviously, but we can do better than that: a drug is a chemical that has an effect on a biological system. Drugs can be derived from natural products (back in the day, virtually all medicines were derived from plants); they might be engineered in a lab using computational chemistry, or they might be existing biological substances, like insulin or adrenaline, for example. Those that act on the brain are often called psychoactive – they change psychological processes such as perception and attention. Drugs rarely have one and only one specific effect: many psychoactive substances also change body systems, increasing or decreasing cardiovascular function, for example. Talking of the 'side effects' of drugs isn't quite the right description. Drugs have all the effects they have, end-of. It's just that some of

DOI: 10.4324/9781003215509-4

the effects are unwanted, and part of the effort in drug design is to try to get as close as possible to a drug that has just one singular effect. It's rarely possible.

Because so much of our focus in biopsychology is on the brain, we need to understand more about basic physiology. We saw neurons in Chapter 1, but we need to look more closely at neurotransmission and, in particular, what happens at synapses. What follows is a very brief review.

NEURONAL COMMUNICATION AND SYNAPTIC TRANSMISSION

Neurons receive chemical inputs from other neurons, usually, though, their dendrites (more on this in a moment). Dendrites send signals into the cell bodies to which they're attached, and there, biological computations are made such that the neuron is able to generate electrical impulses (action potentials aka spikes, which is how they appeared on oscilloscopes during recordings – a sudden spike in electrical activity). Normally, the inside of a neuron has a different electrical charge to the outside, measurable as −70 mV, an internal negative charge (more negatively charged ions inside the neuron than outside). The sudden depolarization – positively charged ions coming in – is what's recorded as a spike of electrical activity.

Action potentials are generated in an all-or-none manner (no half spikes; they either appear or not) at the trigger zone, that part of the axon immediately next to the cell body. Two things to note: first, spikes can come singly or in bursts, clusters of several spikes together, with different patterns having different effects. Second, spikes travel along axons like a wave. Activity at the trigger zone changes the permeability of the axon such that positively charged sodium ions enter through dedicated ion channels. With more positive ions inside the axon, it becomes depolarized (that is, an electrical charge closer to zero; the reverse is hyperpolarization). Activity at the trigger zone changes the permeability of the next bit of the axon so that the spike flows along. The permeability immediately shifts back to normal, with a sodium-potassium pump actively restoring the balance, making sure that the normal levels of sodium, potassium and chloride ions are maintained. This pump is actually an enzyme called Na^+/K^+-ATPase and, because it imports and exports ions against a concentration gradient (that is, from relatively high to relatively low concentrations, the opposite of what would happen naturally), it requires a great deal of energy. So, spikes singly or in bursts travel along axons like waves. Remember, though, that in Chapter 1, we talked about glial cells called oligodendroglia and how they provide myelin to insulate axons such that electrical impulses jump along them (across the nodes of Ranvier) rather than sweep the entire length micron by micron. It speeds transmission hugely. But whether in a myelinated or nonmyelinated axon, what happens when electrical activity reaches the axon's end, the terminals?

At axon terminals, there are boutons connecting points with other neurons where synapses are found. We talk about presynaptic and postsynaptic membranes, presynaptic being the part on the axon terminal and postsynaptic on the receiving neuron. In between is the synaptic cleft – literally, a tiny gap. The whole thing is anchored by minuscule filaments, keeping the connection stable and protecting the synaptic cleft from outside influence. Receptors are embedded in the postsynaptic membrane, molecules that have a lock-and-key relationship with neurotransmitter molecules released from the presynaptic membrane.

How does synaptic transmission work? In the presynaptic terminal, neurotransmitters are gathered into synaptic vesicles, small packets that an axon terminal makes by pinching off a bit of the neuron's membrane. When action potentials arrive at the terminal, they trigger biochemical activity (calcium ions and snare proteins are critical here), which makes vesicles move to the inner surface of the presynaptic membrane, with which they fuse, and in doing so, they rupture and liberate transmitter molecules into the synaptic cleft. The force of vesicles bursting open is sufficient to push transmitters across the cleft and bind to receptors, stimulating cascades of biochemical activity in the receiving neurons. It takes us back to the start: dendrites receiving inputs and passing information to the cell body, which is then able, if necessary, to generate action potentials.

Lots of points still to make: first, neurotransmitters uncouple from receptors and go back into the synaptic cleft. Some of this neurotransmitter content is degraded by enzymes (for example, acetylcholinesterase for acetylcholine, monoamine oxidases for monoamine transmitters), but there are also reuptake molecules embedded in the presynaptic membrane to scoop spare transmitters up, ready for repackaging into vesicles. Manufacturing neurotransmitter molecules is energy expensive, so reuse is the most efficient thing to do. Second, synapses are closed systems with little leakage in or out. However, in parallel with synaptic transmission, neurons can release transmitters into the spaces between neurons. They circulate in this space just like hormones do in the bloodstream, looking for receptor molecules on neurons that aren't enclosed in synapses. This is called variously nonsynaptic or paracrine or volume transmission. Third, the most abundant synapses are between an axon and a dendrite – axodendritic synapses – but there are others, including axosomatic, dendrodendritic and axoaxonic. There are also presynaptic receptors, so that a receiving neuron can give regulatory signals back to a terminal and autoreceptors making negative feedback loops: the released neurotransmitter works on the terminal it was liberated from to regulate further release. Chemical communication at synapses is a dialogue, not a monologue, a complex interplay between pre- and postsynaptic elements. Fourth, neurotransmitters can have excitatory or inhibitory effects on the receiving neuron – that is, they make the generation of action potentials more likely or less. (Technically, we talk about excitatory and inhibitory postsynaptic potentials.) In receiving all sorts of different inputs, dendrites and cell bodies balance positive and negative influences, calculating whether or not to fire action potentials. Some neurotransmitters are also described as neuromodulators, substances whose effects have more to do with modifying the effects of other transmitters rather than driving or inhibiting activity themselves.

Finally, what are neurotransmitters? Not all one thing. In the 1950s, when neurotransmission was first identified, scientists thought that there were just two neurotransmitters, noradrenaline (aka norepinephrine in the USA) and acetylcholine, one to excite and one to inhibit neuronal activity. In the years since, over 100 neurotransmitter molecules have been identified and with them a bewildering variety of receptor types. We can keep this simple by grouping neurotransmitters into four basic classes. There'll be some technical terms in here (catecholamine, for example) that might seem superfluous, but if you get into advanced biopsychology, such terms will come up frequently: it's best to have an acquaintance with them.

(1) Amino acid transmitters are small molecule transmitters. There are a lot of them, but easily the most abundant are glutamate (which is excitatory) and GABA (gamma amino butyric acid – inhibitory). These have very fast actions and operate what one might think of as classic neurotransmission – fast excitation, fast inhibition.

(2) Neuropeptide transmitters. Amino acid transmitters are small molecules; neuropeptides are big, with anything between 3 and 36 amino acids in their structure. They, too, can have excitatory or inhibitory effects but tend to operate over longer periods to provide more sustained activation or inhibition. Many neuropeptide transmitters are found elsewhere in the body, where they have different roles. Leptin and cholecystokinin, for example, are what are called gut-brain peptides because they're found in both places. Some neuropeptides have more mundane names: substance P was first prepared as a powder; neuropeptide Y is so-called because of its relationship with tyrosine, which is identified by the letter Y.

(3) Monoamine transmitters are perhaps the best known: dopamine, noradrenaline and serotonin are primary examples, with histamine another. Dopamine and noradrenaline are in a subgroup of the monoamines called catecholamines; serotonin (aka 5-hydroxytryptamine [5HT]) is a member of a group called indoleamines. More commonly thought of as involved in immune reactions (antihistamines reduce skin irritations and allergies), histamine is an important brain neurotransmitter. Dopamine, noradrenaline and adrenaline are related. Noradrenaline is derived from adrenaline and dopamine is derived from noradrenaline. Acetylcholine is often grouped with the monoamine transmitters, but it is not a monoamine but a biogenic amine. These transmitters are found in very small clusters of neurons whose axons extend everywhere in the brain, having a very wide influence on activity, often neuromodulatory: they enable the work of other neurotransmitters. (And something else: adrenaline and epinephrine are the same thing. In the United Kingdom, adrenaline was first identified

in tissue from the adrenal gland, so adrenaline. More or less simultaneously, it was identified in the United States from tissue in the kidney, which has cells called nephrons: the prefix 'epi' means close to. The prefix 'nor' means that noradrenaline/norepinephrine has the same structure as adrenaline/epinephrine but with a hydrogen atom in place of the N-methyl group.)

(4) Volatile gases are very different. For the other classes of transmitters, you can describe synthetic pathways and storage mechanisms within neurons, processes that enable their release and receptors for them. The volatile gases – principally nitric oxide (NO) but also, somewhat surprisingly given that it's toxic in larger amounts, carbon monoxide (CO) – are synthesized and released on demand, diffusing out of neurons to influence those in close proximity. (CO is a poisonous gas found in car exhaust fumes. In tragic cases in which people died through inhalation, pathologists always used to find higher concentrations of CO in the brain compared to other tissues. It was a mystery resolved by the discovery that CO naturally occurs in the brain. NO is a potent vasodilator – it opens up blood vessels. Sildenafil enhances NO vasodilation.)

Neurotransmitter receptors also come in many different types. A very basic distinction is between ionotropic and metabotropic receptors. Ionotropic receptors work by directly opening an ion channel coupled to it. When a transmitter binds to a receptor like this, particular ions can enter the neuron, where they will have an effect. It could be one of many ions – sodium, potassium, chloride, magnesium or calcium, for instance. Metabotropic receptors, in contrast, do not directly open ion channels but trigger metabolic effects within the neuron, leading to cascades of biochemical activity. Substances called G proteins (guanine nucleotide-binding proteins) are a key part of this, molecular switches inside cells mediating inputs from outside. Individual neurotransmitters can act on multiple types of receptors. For example, dopamine has different receptors numbered one to five (D1–D5). Noradrenaline has alpha-1, alpha-2 and beta receptors. Glutamate has NMDA (N-methyl-D-aspartate), AMPA (alpha-amino-3-hydroxy-5-methyl-4-isoxazolepropionic acid) and kainate receptors – all ionotropic – and GluR3 metabotropic receptors. Acetylcholine has muscarinic and nicotinic receptors, named for the substances first found to uniquely affect them. Muscarine is the toxin in fly agaric mushrooms (the red spotty ones, very poisonous), and nicotine is, well, nicotine. (Nicotine and the tobacco plant [*Nicotiana tabacum*] are named after Jean Nicot (1530–1604), French ambassador to Portugal who sent the first tobacco seeds to France. Tobacco has the highest levels of nicotine, but potatoes, tomatoes and aubergines (eggplants), among others, also have traces.)

We've tried to keep this simple but can't disguise the fact that neurotransmission is phenomenally complex, a continuous process going on in the brain, the peripheral nervous system and, of course, between nerve and muscle (where neuromuscular junctions, specialized synapses, use the neurotransmitter acetylcholine to control muscle fibers). It almost defies belief: billions of neurons whose biochemical activity is in constant flux, signaling to each other in order that movement, perception, attention, memory and all the rest of psychological life can proceed as normal. Rough estimates suggest that a single human brain neuron will make about 1,500 synaptic contacts. Do the math: 86 billion human brain neurons (86,000,000,000) times 1,500 synapses each equals 129 trillion synapses (129,000,000,000,000). The probability of a human brain ever being in the same state twice is infinitesimally small.

Now, where can drugs act on all of this? As you've no doubt realized, the answer is many places. We can list them as follows:

- Receptors – presynaptic, postsynaptic, autoreceptors: at any of these drugs can act as agonists (mimicking the natural ligand – which is what we call any molecule that binds to a receptor) or antagonists (blocking the receptor from receiving other inputs). This can get complicated quickly: there are partial agonists that, even when they've occupied all the receptors they can, don't have the full capability of the natural ligand. Inverse agonists seem slightly weird: they bind as agonists do but have the opposite effect (unlike an antagonist, which is just a block – indeed, they're often called receptor blockers). Partial inverse agonists combine the two effects.
- Drugs can promote or inhibit the release of neurotransmitters, while others can block the transmitter synthesis. Some drugs are given therapeutically to enhance synthesis. L-DOPA (aka levodopa or l-3,4-dihydroxyphenylalanine), the naturally

occurring precursor of dopamine, has for decades been given to Parkinsonian patients in an effort to restore massively depleted levels of this transmitter.

- There are drugs that promote or block the effects of enzymes that degrade neurotransmitters in the synaptic cleft – blocking enzymatic degradation can prolong the effect of a naturally occurring transmitter. In a similar manner, there are drugs that can block or enhance reuptake mechanisms.
- There are drugs that are toxic for particular types of neurons – neurotoxins. In experimental neuroscience, the best known is 6-hydroxydopamine (6-OHDA), which can enter dopamine and noradrenaline neurons through their reuptake molecules. Once inside the neuron, 6-OHDA breaks down and produces hydrogen peroxide, which is fatal to the cell. Excitotoxins, like ibotenic acid (aka ibotenate) are structural analogues of glutamate, which, when injected into the brain, bind to glutamate receptors, causing the uncontrolled passage of ions and cell death. (While injectable neurotoxins like all of these were once used in experiments to damage neurons in particular places or of a particular type, they are less commonly used in research now. Genetically modified mice can be created that do not have certain neural elements, or optogenetics can be used to prevent the activity of groups of neurons. Molecular biological techniques have had a major impact on how scientists approach questions like, "What would happen if these specific neurons were inactivated or not present?")
- And there are drugs that interfere with transmission along axons. Tetrodotoxin blocks sodium ion channels, effectively stopping action potentials. The name comes from *Tetraodontiformes*, types of fish. Pufferfish (aka blowfish) are the best-known example. Considered a delicacy in some countries, they have lethal levels of tetrodotoxin in their liver and sex organs: best not to eat them. (Fans will know: "One Fish, Two Fish, Blowfish, Blue Fish" is episode 11, series 2, of 'The Simpsons'. Homer tries sushi and believes, wrongly, he's been poisoned because of incorrect blowfish preparation.)

We could go into more detailed mechanisms and include agents that can affect gene transcription. We believe, though, that we've made our point: synaptic transmission is incredibly complex, offering multiple points at which drugs can act. Two last things: there are multiple routes of drug administration. Experimental neuroscientists working with lab animals (which are called *in vivo* experiments) can deliver drugs in microliter quantities direct into the brain or use optogenetics – we discussed this in Chapter 1. However, for therapeutic and recreational drugs, such direct intracranial approaches are inappropriate except in very rare instances of neurological disease. Instead, there are many techniques for delivering drugs, including injections (into the bloodstream, muscle, the peritoneal cavity [into the body] or other places), inhalation as gas or smoke and by swallowing, ingestion into the stomach. You need to keep in mind that the blood-brain barrier (we talked about this in Chapter 1) works to limit access into the brain: small molecules can pass through, but not large ones. Psychopharmacologists, like any other kind of pharmacologist, also need to recognize how drugs are distributed around the body – pharmacokinetics – and the principles of ADMET – that is, administration, distribution, metabolism, excretion and toxicity. It's important to know that any drug being given gets to where it's supposed to (not digested in the gut, for example), acts where it should, has its metabolic products eliminated and is not toxic in any way. Bioavailability is an important concept within pharmacokinetics: when a given dose of a drug is given, how much reaches the target? The kinetics, ADMET and bioavailability all have to be optimized in developing drugs of any sort. In parallel, we need to understand pharmacodynamics. While pharmacokinetics describes how a pharmacological agent moves around the body (anything with the pre- or suffix kinesis/kinetic has to do with movement), pharmacodynamics describes the body's biological response. In the simplest terms, pharmacokinetics (PK) is how a drug gets around the body, and pharmacodynamics (PD) is how a body reacts to a drug.

PSYCHOPHARMACOLOGY – BEHAVIOR, PHYSIOLOGY AND DRUG ADMINISTRATION

Working now and over the past centuries, there are thousands and thousands of scientists, most of whom remain anonymous to the wider public. A very small number become household names – Darwin, Einstein – and some become well-known within their fields, their reputations lasting longer than the usual half-life of fame. Peter Dews was a scientist who deserves to have his reputation maintained. He was a pharmacologist from Yorkshire who, after graduation and a series of research positions in the United Kingdom and the United States, was appointed at Harvard Medical School. He already had an interest in the effects of drugs on the brain and behavior, but at Harvard, he met B. F. Skinner (whose work was discussed in the Introduction). Peter Dews had the imagination to see beyond the traditional confines of pharmacology. By combining the administration of drugs with Skinner's analytical approach to behavior, he made studying the behavioral effects of drugs objective and quantitative. Because his focus was on behavior rather than psychological processes, the field he developed was more accurately called behavioral pharmacology, but the principles of his work hold good for psychology. Studies of lab animals working in Skinner boxes have produced valuable insight into the effects of drugs on learning, memory attention, decision-making and a raft of other psychological processes. Peter Dews was first-in-field and deserves thanks and respect.

We could approach psychopharmacology with a list of drugs and what they do – morphine does this, amphetamine does that and so on – but in this basic book, we're going to focus on principles. There is a huge point of convergence between biology and psychology, with their interactions being far from straightforward. How drugs act to shape psychological processes and behavior is not a simple one-to-one but a genuine interaction. For example, drugs might have what are called rate-dependent effects, described by Peter Dews and other researchers, such as Mel Lyon and Trevor Robbins. Rate dependency means that a drug's effects on a given behavior depend on what the baseline rate of activity was before administration. Amphetamines are a classic case. An activity normally occurring at a high rate is likely to be suppressed by amphetamine, but another with a low baseline rate can be increased by the same dose of it. The effects of the drug are not simply a function of physiology and pharmacology but depend on current active behavior as well. What an animal or person is doing affects the response to drugs, as does their present physiological state. In recent years, this has been highlighted not just by rate dependency but also by the emergence of a new field: chronopharmacology.

Circadian rhythms are fluctuations that occur within a period of approximately 24 hours – circadian comes from *circa dies*, about a day. Left to run on their own, circadian rhythms will drift away from a 24-hour cycle, but as long as individuals are operating in the natural world – sun comes up, sun goes down – they are entrained to a matching 24-hour rhythmicity. Within the brain, the suprachiasmatic nucleus of the hypothalamus was identified as a key controller, setting a rhythm whose mechanism was based on the activity of specific genes called CLOCK genes. (Suprachiasmatic means above the optic chiasm, the point immediately below the brain where the optic nerves cross, information from the right side of each eye going to the left hemisphere and *vice versa*. The suprachiasmatic nucleus taps into the visual information coming into the brain, the first place to access it.) The suprachiasmatic nucleus isn't the sole owner of circadian rhythmicity. There are very many rhythm generators around the body. The suprachiasmatic nucleus is special because it is a master controller of multiple separate processes around the body. Pharmacokinetics and pharmacodynamics are both affected by circadian rhythms that determine what sort of state a body is in at a given time. Diverse biopsychological events like pain, blood pressure and both psychological and physical stress are subject to circadian variations that affect how drugs work. Age, biological sex and ethnicity also come into play here. Altogether, it means that the frequency and timing of administration will affect how the drug acts and what sort of physiological and behavioral responses will come of it.

Everyone is probably aware of other processes to do with psychopharmacology and drug administration: dependency and tolerance. We'll deal with these through the context that most people have for them: drug addiction.

PSYCHOPHARMACOLOGY – DRUG ADDICTION

What is drug addiction? A state in which the presence of a drug is required for a person to function normally, though, of course, the idea of normally can spiral downwards rapidly to the point where life is anything but normal. Addiction – whether to legal or illegal drugs – is likely now to be referred to as substance abuse or substance abuse disorder. Using substance abuse helps distinguish drug addiction from other serious forms of addiction, such as gambling, and is more inclusive. Alcohol might not be popularly thought of as a drug, but it is most certainly a substance and most certainly has addictive properties.

In the definition, saying that a drug 'is required' carries a number of important concepts, such as craving, which is an unnaturally strong desire to obtain and use a drug (or indeed something else – carbohydrate craving has been described in depression and other disorders). Likewise, there is dependence. Physical dependence is a state in which the presence of a drug is necessary for a body to function: unpleasant physical withdrawal symptoms occur if the drug is unavailable. However, more often than not, what's meant when people talk about dependence is psychological dependence, evidenced by impairments in physical and mental health. Important features used diagnostically include spending more and more time with drugs (getting them, using them, recovering after use), neglecting responsibilities, damaging interpersonal relationships, giving up other more desirable activities and engaging in risky behaviors leading to dangerous situations. (These feature in the detailed criteria presented by the Diagnostic and Statistical Manual of Mental Disorders of the American Psychiatric Association – DSM-5 [the fifth edition]). Tolerance is a concept that sits next to dependence: it's a reduced reaction to a drug following its repeated administration. One sees tolerance developing when the original effect of a drug can only be repeated by giving larger and larger doses. Tolerance can appear physically and psychologically, and perhaps surprisingly – and then again perhaps not – it can develop in one without the other. For example, chlordiazepoxide (aka Librium), a classic benzodiazepine, has both sedative and anxiolytic properties – a physical state of sleepiness and, psychologically, reduced anxiety. Librium's sedative property shows tolerance; its anxiolytic one does not. The opposite of tolerance is sensitization (aka reverse tolerance): less drug is needed on repeat administration to achieve the original effect. (And something to note: anything ending in -lytic means reduced; the opposite suffix is -genic. So, for example, anxiolytic reduces anxiety; something anxiogenic increases it.)

Drugs clearly raise social and cultural issues: illegal drugs not only compromise users; they have horrible consequences through the criminal activity that supplies them. When we talk about drugs in the context of addiction, we typically mean things like alcohol, amphetamines, barbiturates, benzodiazepines, cannabis, cocaine, hallucinogens (mescaline, peyote and LSD, for example), nicotine and opiates. In most Western countries, some of these are legal, some illegal; some have medical value, some don't; some stimulate, some sedate; and some form an enjoyable part of very many people's lives, only becoming problematic in a proportion of cases. Our inventory of drugs isn't exhaustive: khat, for example, rarely appears in textbook lists. Addictive and with no medical purpose, khat is a widely used stimulant derived from the *Catha edulis* plant grown in East Africa. Chewing its buds and leaves liberates cathinone to create a state of euphoria and stimulation. It's illegal in most Western countries.

Once upon a time, addiction was simply thought of as a weakness of will, a giving way to pleasurable habits, a simplistic way to account for the problem by pointing at the feebleness of the addict. Things have moved on. We recognize now that drugs of abuse – legal or illegal, of medical value or not – act on the brain to produce effects, and as we just grasped by looking at the complexity of neurotransmission, psychoactive drugs can change brains through a whole raft of mechanisms. Brain changes are thought to reflect the transition that people go from a careful and controlled use of drugs to a compulsive addiction. Is there an inevitability to this? It's important to recognize the dangers, but equally, it's important to understand that drug use need not unavoidably lead to addiction. In a post-war report on drug use by US Army personnel serving in Vietnam, it was recognized that

> [a]lmost half of the representative sample had used heroin or opium while in Vietnam. The majority of these had developed neither physiological nor psychological dependence. In the year after their return, about 10 percent had some experience with opiates, but less than 1 percent showed signs of dependence. [3]

How might addiction come about? An old view was not so much a matter of will – that really is a very dated approach – but that the unpleasant effects of withdrawal from a pleasurable drug caused it: addiction as the avoidance of unpleasantness. More recent approaches suggest that either learning goes wrong, with stimulus-response habits becoming so strong as to be unavoidable (think of it as a see-drug-use-it compulsion), or decision-making becomes impaired, leading to impulsive judgments (a sort of can't-help-myself). These could both involve changes to neural circuits in the brain – largely through the frontal cortex and basal ganglia (Chapter 1) – known to be involved in learning and decision-making. A popular theory, and a more nuanced one, is the incentive-sensitization approach, which argues for dysregulation of those brain systems that attribute incentive salience to events – desired things (incentive) becoming more prominent (salient) to an individual – causing compulsive motivation. This is part of a more general theory of 'wanting' and 'liking' developed by Kent Berridge and Terry Robinson. Wanting and liking aren't the same, either conceptually or as brain processes. Wanting equates to the motivation to get something and is separate from liking the enjoyment of it. Wanting, but not liking, is thought to be dependent on dopamine activity in the nucleus accumbens (see Chapter 1). The incentive-sensitization approach to addiction has at its heart the idea that wanting is amplified unnaturally, with a basis in long-lasting changes in dopamine activity present in some individuals but not others. It's a neural model that highlights addicts' need for a drug without their necessarily finding pleasure in it. Is this the answer? It's not a fixed certainty, but changes in brain mechanisms to do with learning, wanting and decision-making seem entirely plausible and consistent with a lot of other data.

It's entirely possible that all these sorts of mechanisms come into play, not necessarily all in the same person but overall. What begins with liking – the pleasure drugs can bring – gives way to cravings, habits, compulsions and a distorted sense of wanting. One thing that is clear is that learning has a part to play in this. One of the reasons why it can be so hard to quit is that many things become associated by conditioning with drug use. I. P. Pavlov's work on conditioning is important here: just as the sound of the bell made dogs salivate in anticipation of food, drug paraphernalia has a conditioning effect: the sight of them will trigger feelings of need. Given that drug paraphernalia includes a lot of everyday objects – lighters, cigarettes, spoons, tinfoil, let alone needles and syringes – the presence of cues can be hard to overcome. It gets worse. The place in which drugs are used – whether the street, a particular building or rooms within it – has a conditioning effect too. These effects are not simply behavioral but physiological as well, preparatory responses that get the body ready for the shock of drug taking, tolerance mediated by conditioning. There is another really bad side to this. Death by overdose can be the result of straightforwardly taking too much, but it isn't always the case. Drug-experienced addicts who have fatally overdosed can be found to have no more drug in their bodies than they would normally have used. A cogent argument made by Shepard Siegel is that overdose can result from taking a 'normal' amount of drug but in a new and different environment, one that does not have the conditioning effect on the body. Unprepared, the 'normal' dose is lethal. This is biology and psychology intimately linked, but in a very sad way.

PSYCHOPHARMACOLOGY – THERAPEUTICS AND THE BRAIN

Therapeutic drugs are given to people who have, in the broadest sense, two different kinds of illness. One is organic brain damage in which there is clear physical evidence of a problem, whether degenerative like Alzheimer's, ischemic (caused by obstruction of the blood supply, as in stroke) or mechanical, head trauma caused by accidents, falls leading to abrupt contact with the floor or any number of unfortunate events. The other kind of illness is psychiatric, in which the brain pathology is much less obvious, but the signs and symptoms (what a clinician can see and what a person reports) are real and apparent. We will talk more about pathologies of both sorts in Chapter 7. For now, we want briefly to say something about drug treatments.

Ischemic brain damage – stroke – can happen when a blood clot forms, the clot causing a blockage that restricts the flow of blood. The initial treatment is aspirin, which is a painkiller and has antiplatelet properties, blood platelets being cell fragments that help form clots, which after a stroke you don't want. Later, other antiplatelet drugs might be used (for example, clopidogrel and dipyridamole, as well as anticoagulants, to reduce the probability of more clots being created; Warfarin is the most common, but there are others; heparins, anticoagulants used in other medical contexts are less after stroke used because they have to be given by injection). The point is to manage the underlying case – blockage in the bloodstream – and prevent further problems, but these aren't therapies directed at brain tissue *per se*. In the case of neurodegenerative diseases, brain cells and their activities are the target, not the blood supply. The best-known case is that of Parkinson's disease. Once it became clear in the early 1960s that this was associated with the loss of dopamine from the basal ganglia, treatments focused on replacement therapy to get more dopamine into the brain. This was achieved by giving L-DOPA (aka levodopa or l-3,4-dihydroxyphenylalanine), a precursor of dopamine that can be taken up and used by neurons. While it cannot restore impulse-dependent dopamine release (that is, the release of dopamine from terminals driven by action potentials), it does make dopamine available where it's needed. In the first trials, too much was given because of problems with it being metabolized in the gut, but that problem was solved by creating formulations that could more easily get into the brain without being degraded. The development of L-DOPA therapy was a triumph. It's not a cure, and because it doesn't stop the progression of the disease, there comes a point for every patient when it stops being effective. But what it does brilliantly is alleviate the symptoms, giving years of benefit that would otherwise be unavailable. (As an aside here, Oliver Sacks's book 'Awakenings' gave a remarkable insight into the early trialing of L-DOPA. In 1990, Penny Marshall made the book into a film starring Robert De Niro and Robin Williams. Poignant and still worth watching.)

There was a point some years ago when the loss of the neurotransmitter acetylcholine from the cortex was thought to be the key problem in Alzheimer's. At the time, a similar replacement therapy approach was tried. Regrettably, drugs promoting acetylcholine availability had scant effect, the issue being that loss of acetylcholine-containing neurons was a consequence of the disease and not the cause of Alzheimer's. The key features of Alzheimer's disease are the formation of plaques and neurofibrillary tangles: plaques are accumulations of amyloid protein, and tangles are the knotted remains of dead fibers. If the early formation of these could be identified and drugs used to stop their creation it should be beneficial. This is where current research is focused – disease-modifying drugs to slow progression. Drugs that act through the immune system (Chapter 2) are being developed in an effort to clear amyloid protein. Clearing amyloid through immunotherapy should, in theory, slow the progression of the disease. Donanemab, produced by the pharmaceutical company Eli Lilly, is having success in trials. This company also has Remternetug, a second generation immunotherapy that targets amyloid in a different way. Lecanemab (aka Leqembi) from the pharmaceutical company Eisai is another immunotherapy drug. It seems to slow the progress of the disease, and in January 2023, it was approved for use in the United States by the Food and Drug Administration; it's currently (spring 2023) under consideration in the United Kingdom and European Union. These are a long way from a cure, but undoubtedly, there is cause for optimism in the treatment of a truly unpleasant neurodegenerative disorder.

Strokes and neurodegenerative diseases are unambiguously brain disorders. There are pathologies that can be identified that interfere directly with the workings of the brain and have consequences for behavior and mental activity. We think of these as organic diseases treated by neurologists. With psychiatric disorders, however, pathology in the brain is much less apparent or even absent. The signs and symptoms are clear and distressing, but while there might be assumptions about associated brain states, there is not the same clarity that exists with neurological disorders. (We will talk more about psychopathologies in Chapter 7.)

Table 3.1 lists some of the commonly used drugs in combating psychiatric and neurological disorders. An immediate thing to notice is how involved the monoamine transmitters are in this – dopamine, noradrenaline, serotonin, histamine. These are the transmitters used by the neurons that make up the ascending reticular activating system, that collection of neurons through the brainstem whose axons spread to every part of the brain (see Chapter 1). One implication of this is that many of the drugs in current use have actions that are widespread, not targeted. As a corollary to this, it's worth noting that antipsychotics, antidepressants and anxiolytics tend not to work immediately but take up to two weeks to have effects, considerably longer than the few hours it takes to become distributed through tissues. Whatever effect they have, it's not just a matter of changing a brain system and altering psychology and behavior instantly.

The first antipsychotic drug was chlorpromazine, from the phenothiazine group of compounds and available under the name

Table 3.1 Common pharmaceutical treatments for psychiatric and neurological disorders and their effects on brain neurotransmitter systems

Condition	Prescribed drugs	Neurotransmitter systems
	These are the principal prescribed drugs; others are available, acting through much of the same systems	Actions include receptor blockade/stimulation, reuptake blockade, interference with enzymatic destruction or on synthetic pathways
Schizophrenia and other psychoses	Aripiprazole, Olanzapine, Risperidone	Dopamine: some effects on serotonin, histamine, acetylcholine
Depression	Fluoxetine, Citalopram, Venlafaxine, Mirtazapine	Serotonin and noradrenaline: some effects also in panic and obsessive-compulsive disorder
Bipolar disorder	Priadel, Camcolit, Liskonium, Li-Liquid – these all contain Lithium (Li)	Bipolar disorder is not best treated with antidepressants. Lithium is preferred. Antipsychotics may be needed in mania
Anxiety	Fluoxetine, Buspirone, Propranolol, (Diazepam)	Serotonin, noradrenaline, adrenaline (GABA)
Parkinson's and related disorders	Levodopa, Carbidopa Entacapone, Selegiline, Ropinirol, Pramipexol	Dopamine: some effects on noradrenaline
Alzheimer type dementia	Donepezil, Memantine	Acetylcholine

Largactil. In the early 1950s, it was used to treat vomiting, pain, nausea and convulsions. It was also used to treat shock in serving soldiers, who reported disinterest in events following chlorpromazine administration. It was only then that Henri Laborit, a French surgeon and neurobiologist, used it to treat mania and, after that, schizophrenia. It blocks dopamine receptors, but in the 1950s, no one knew that – indeed, at this point, dopamine was not even recognized as a neurotransmitter. Understanding the actions of chlorpromazine and developing new and better products to block dopamine activity – central to antipsychotic action – have produced better treatments, but it all rests on chlorpromazine being used on a trial-and-error basis. Do antipsychotic drugs work? Yes. Are they just sedatives – major tranquilizers as they were once called? No, they do appear to have genuine benefits. Do they cure schizophrenia and related disorders? No. They're palliative, helping manage the disease and delay a second schizophrenic episode after a first one has been treated. Antidepressants are prescribed widely but have the same sort of history: accidental discoveries rather than principled. In the 1950s, tricyclic antidepressants like imipramine were developed, actually from chlorpromazine. Lacking antipsychotic effects, they were found to have benefits against depression. The newer serotonin and/or noradrenaline reuptake blockers (SSRIs) were developments from the old-school tricyclics, minor tranquilizers. Do they work? Again, yes. Are they a cure? No – they're also palliative and can help patients into a better state from where psychological therapies – counseling and cognitive behavior therapy, for example – can have a better effect.

The point of looking at psychopharmacology through the lenses of addiction and therapeutic regimes has been to highlight how biology and psychology intersect through the effects drugs have on neurotransmission. How brains work, how neurological and psychiatric conditions come about, how we can affect treatment – all of this is fundamental and deeply practical biopsychology. What we'll do in the next chapters is explore more about how normal psychological processes work and what biological substrates they have.

CHAPTER 3: SUMMARY

- *The human brain contains some 86 billion neurons. Each neuron makes approximately 1,500 synaptic contacts, making a likely total number of synapses of around 129 trillion. Understanding the nature of synaptic transmission is a fundamental goal of neuroscience.*
- *Where and how do drugs (whether recreational or therapeutic) act in the brain? There are many ways: they can act at receptors to block or stimulate activity, change reuptake or transmitter degradation, or affect the release of transmitters, and they can act on axons to affect the flow of electrical activity.*
- *Psychopharmacology studies the relationships between behavior, psychology, physiology and drug administration. How drugs are given and how they are distributed through the body is important for understanding their effects.*

- *A major part of research in psychopharmacology has to do with substance abuse (drug addiction). Understanding the interactions between the brain, body and drugs helps one understand factors such as craving, dependence, withdrawal, tolerance and sensitization. There are multiple theories of addiction, many based on conditioning or on processes such as incentive motivation.*
- *Psychopharmacology also has a major interest in therapeutic drugs, their routes of delivery, targets and effectiveness. The development of new pharmaceuticals has often been through fortuitous discovery, but there are newer developments that are based on better theories – immunotherapies, for example.*

DEVELOPMENT

It seems an odd way to start, but this is what this chapter is not about: evolution. When biopsychologists talk about development, they can mean the evolutionary development of a species over extended time or the development of an individual over a lifespan. We're just going to focus on how biopsychology deals with the physical and psychological development of the individual, covering questions like how does a brain grow, how do psychological qualities develop, to what extent do these processes continue through the lifespan and what happens in aging?

We also need to include something about genetics and genomics. Genetics concerns inheritance and variation in living organisms; genomics is the study of DNA, whether the object of investigation is all the DNA in a species, in a person or in cells. Regrettably, over many years, there have been polarized debates about race and genetics, notably in regard to intelligence. The National Human Genome Research Institute, part of the National Institutes of Health in the United States, makes this important statement, which we endorse.

> Race is a social construct used to group people. Race was constructed as a hierarchal human-grouping system, generating racial classifications to identify, distinguish and marginalize some groups across nations, regions and the world. Race divides human populations into groups often based on physical appearance, social factors and cultural backgrounds.

The idea of different human 'races' was first crystalized by Linnaeus (Carl von Linné, 1707–1778. It was the style at the time to Latinize your name if you were a scientist). Linnaeus is highly regarded as the father of systematic biological classification, but in the case of people, he got it badly wrong (as many contemporaries told him). People are people; that's the end of it: *Homo sapiens*, one species, one people. Having said all that, let's start with genes.

GENES, GENETICS AND GENOMICS

Deoxyribonucleic acid: DNA has to be the most familiar abbreviated acronym in the world. All too often, it's used inappropriately. Sports commentators love to say things about a football club's DNA – how a style of play is "in their DNA" – when they don't mean DNA at all but culture. So let's keep this tight. DNA enables species, through reproduction, to change. For any organism, its DNA is a mix of that from both parents, a blending which means that adaptation, mutation and evolution can happen. Molecular biology gets into all aspects of the life sciences, including biopsychology, so understanding something about how genes work is worth it. We'll go step-by-step through some of the basic things you ought to know.

Amino acids and proteins. Proteins are made in the human body from twenty different amino acids, ten of which are already in our bodies and ten of which (essential amino acids) come from what we consume. Proteins are built from exact sequences of amino acids; we call this protein synthesis. Proteins construct body parts – our actual structure – as well as, for example, chemical controllers like hormones and enzymes. DNA and RNA control protein synthesis. A protein's shape determines its function.

DNA – deoxyribonucleic acid – is a molecule found in every cell of your body. DNA has two strands in a double helix. (Imagine a twisted ladder, two sides spiraling round each other, joined by rungs.) The strands are connected to each other by chemicals called nucleotides, present in what are called base pairs. There are four nucleotides defined by their bases: cytosine, guanine, adenine and thymine. In making base pairs, adenine is paired with thymine, cytosine with guanine. If the two strands of DNA are separated, they won't be the same because one of each base pair stays with each strand. Think of the strands as photographic positives and negatives, the same but different.

RNA is ribonucleic acid, one strand, not two. Messenger RNA (mRNA – made inside cell nuclei) is a copy of part of a single strand of DNA. mRNA in turn makes transfer RNA (tRNA), and tRNA uses the code copied from DNA to drive protein synthesis. There are other types of RNA (we'll find one in a moment), but these two are the most important.

The genetic code. A gene is a sequence within the whole DNA molecule. Each one is defined along the DNA molecule by codons. These short segments work like punctuation, showing where each gene is: there are start codons and stop codons that bracket a gene. When we talk about a genetic code, we're talking about how the four bases, cytosine, guanine, adenine and thymine, are ordered. This code means that different proteins can be made through transcription and translation. Transcription is making mRNA from DNA; translation is the process of tRNA using the code to make proteins.

Chromosomes are found in the nucleus of every cell and contain all the cells' DNA. Chromosomes come in pairs: humans have 23 pairs (so 46 chromosomes per cell) except for sperm and egg cells (aka together as gametes or reproductive cells), which have one chromosome each, so when sperm fertilizes an egg, it creates a new cell with two. Human chromosomes differ in size – chromosome 1 for instance is three times the size of chromosome 22. Chromosome 23 can be X or Y: biological females are XX and biological males XY. (X and Y refer to the shape of the chromosome.) In biological females, one of the X chromosomes in each cell is permanently suppressed, a process called Lyonization after the English geneticist Mary Lyon, who discovered this. There are genes associated with various disorders located in different human chromosomes. These include neurodevelopmental disorders, such as Down syndrome (trisomy 21 – an extra copy of chromosome 21), Fragile X and Rett syndrome, which are both X chromosome associated.

How precise is all this? Less than might be thought. Alleles are alternative versions of a single gene: one allele will be dominant, one recessive. Having blue or brown eyes depends on different alleles of the same gene. More than this, variation in the genome can come as a single nucleotide variant (SNV) when a single nucleotide in the genome sequence is altered; as a single nucleotide polymorphisms (SNP; pronounced 'snip'), which involves a single base pair with a substitution in it – for instance, thymine/adenine for cytosine/guanine or *vice versa*; or copy number variations (CNV), which is a general term given to alterations (including duplications, deletions, translocations or other states) that produce too many copies of a gene; or as a single gene mutation that affects just one gene. Gene repair is also possible. Radiation or cancers, for example, can damage DNA, and if the damage isn't too great, it can be repaired by DNA repair molecules. The point isn't to blind you with science – the terminology of genomics is intimidating for nonspecialists – but to highlight the fact that the genome is not quite the fixed thing that it's imagined to be.

Gene disposal happens every day. Gut enzymes break down RNA and DNA in the material you consume. Ribonuclease (for RNA) is made in the pancreas and deoxyribonuclease (for DNA) in the small intestine. RNA and DNA in the protein that you eat, whether plant, animal or fungus, are completely broken down, and the bits and pieces used by your body.

At the heart of all this is how genes code for the structure and manufacture of proteins. It was known for a long time that this coding DNA was only a modest part of a whole DNA molecule. The rest of it was referred to as junk DNA. It's hard to understand why anyone would be so dismissive as to call a large part of a complex molecule essential for life junk. It isn't. We now talk about it as noncoding DNA, and it brings us to epigenetics.

Are genes active all the time? As a whole genome, yes, but individual genes can be switched on or off. Gene expression is when a gene is switched on and working. Some genes are expressed all the time, but others may be needed only at a critical point in development or in special circumstances. Gene silencing is when a gene is switched off. Genes are actively regulated all the time by controlling genes that manufacture proteins that turn other genes on or off.

Gene expression and gene silencing are examples of epigenetics, changes in how DNA is read. In these cases, the structure of DNA is unaltered, but its activity is changed. Physically, epigenetic effects are brought about in various ways, including through the activity of sRNA – small RNA – which is able to select and deactivate small portions of the genome. Epigenetic changes can occur for a variety of reasons, such as events in the womb before birth; influence of environmental agents, including drugs and chemicals; and through exercise, aging and disease. These changes are often reversible. In a remarkable study, NASA found that astronauts, after a year in space, had marked epigenetic changes in their DNA, but it returned to preflight levels within a month of returning to Earth. What's even more of a surprise is that epigenetic changes can be heritable (passed on to offspring), which flies in the face of what has been received wisdom for generations. The heritable changes are those that happen in sperm and egg cells. Many of the epigenetic effects that happen in these cells are not permanent, cells changing and then returning to their original state. But some epigenetic changes in the gametes are permanent, meaning that they can be passed across generations.

[An historical note: in the late 1920s the Russian geneticist Trofim Lysenko claimed that characteristics acquired during life could be passed on to subsequent generations. He believed that changes in an organism brought about by environmental effects were the primary drivers of inheritance. His ideas went far beyond the limited range of heritable changes brought about through epigenetics. Lysenkoism, as it became known, was driven by political ideology not scientific rigor. It was an attempt to validate the belief that the organization of society is the driver of change in people, not evolutionary biology. Lysenkoism was adopted for many years in the USSR and for a while in the People's Republic of China.]Genes are there in every aspect of biology and effect the development and operation of it all. We'll concentrate now on brain development, in which both genes and environmental events have crucial roles.

BRAIN DEVELOPMENT

In Chapter 1, we had an extended discussion of how brains develop, but we didn't talk so much about the control genes have over this. The development and connectivity of the brain is dependent on families of related genes. One of the earliest to have an effect is Sonic Hedgehog. Really. Not the animated one, but a gene called sonic hedgehog (SHH). There are several hedgehog genes (including DHH, the desert hedgehog and IHH, the Indian hedgehog) given the collective name 'hedgehog' because if they don't work properly, embryos become deformed and spiky-looking. Sonic is quick off the mark. In the fifth week of development, human embryos develop three cell layers: endoderm, mesoderm and ectoderm (inside, middle and outside, respectively). Sonic hedgehog stimulates the ectoderm to begin the process of neurulation, growing the nervous system. As we discussed before, what follows is the creation of the neural plate and then the neural tube, with neurons developing from radial glial cells and neural stem cells.

The coding genome is central to brain development. Proteins are literally the building blocks of the nervous system and work to guide development. Epigenetic effects from the noncoding genome are also present, sometimes bad, sometimes good. For example, dietary deficiencies in a pregnant mother can lead to spina bifida. There are several types of this, all involving malformation of the neural tube. Making sure that the fetus is properly fed with all the right nutrients and vitamins – especially folic acid (vitamin B9) – minimizes the risk of spina bifida. Conversely, rather than a deficiency, excessive alcohol intake during pregnancy can lead to fetal alcohol syndrome, in which neurons mis-migrate, deforming the brain. These are bad epigenetic effects. An example of a good one is the stress reduction in infants that comes from breastfeeding.

Human brains are scaled-up primate brains. What makes the human brain physically different is not possession of an extra part. We have the same structures as pretty much all other mammals. Size is the big difference. The human brain has around 86 billion neurons, three times the number of any other primate. The human coding genome is very close to that of other primates, with 99% similarity, though this is put into perspective when you realize that human DNA is 70% the same as a zebrafish (*Danio rerio*, a freshwater minnow from south Asia) and 60% the same as a banana. What these similarities emphasize is that a lot of the proteins of which plants and animals are made serve common cellular functions and as such are very widespread. However, it's really important to know that there are a handful of genes unique to humans. They don't magically code for something like language (which people did once think) but enable an enlargement in the numbers of neurons across all structures, their connectivity, the subtleties of their structure and the degree of cortical folding. Cortical folding is the creation of sulci and gyri, the creases and ridges that allow a greater volume of material to be crammed into a small space. How the human brain develops is essentially the same as that of other mammals. So let's ask a series of questions bound into one big one: what does the time course of human brain development look like?

How many neurons does a newborn have? When does neurogenesis end? A newborn human infant already has the vast majority of its neurons but the prefrontal cortex and the cerebellum continue to develop neurons postnatally – a process called neurogenesis. Although the volume of postnatal neurogenesis is small, their brains are not yet anything like fully connected or ready to work. Until recently, it was thought that adult brains could not create new neurons, but as we noted in Chapter 1, it is possible. The hippocampus certainly can do this, and there may be other places too.

When is myelination complete? This takes a while. Myelination happens from the back of the brain to the front (aka caudal to rostral), from top to bottom (aka dorsal to ventral) and from the center to the outside (aka medial to lateral). We discussed these directional terms in Chapter 1. It means that the brainstem is first to be myelinated and within the cortex, the occipital lobe at the back becomes myelinated before the frontal regions. Much of the myelin in the human brain (provided by glial cells called oligodendroglia) is in place by about 2 years old, but that is an incomplete story. The prefrontal cortex continues to myelinate into adolescence, with its progress altered by environmental and gender-related factors. So the notion that there is a time "when myelination is complete" is not quite right. There is increasing evidence that myelination is activity dependent – that is, the activity of neurons influences the degree to which they become myelinated. Other recent research highlights that the cortex can produce new oligodendroglia and myelin even into adulthood. Once neurons have been insulated by myelin, it remains stable, but if it is disturbed by disease or injury, some remodeling of myelin may, in some cases, be possible. Recent discoveries about the adaptability and continual recharging of myelin open up new avenues for important research. Multiple sclerosis is a disease that has multiple forms, from slowly progressing and mild to quick and devastating, but all involve de-myelination. Understanding myelin repair and re-myelination is crucial if this terrible disease is to be dealt with, palliatively or with a cure.

When does synaptogenesis end and when does synaptic pruning happen? In a developing human brain, the axons of neurons grow away from the cell body in search of targets. This process is guided by signaling molecules that make sure the broad outline of brain wiring is correct – that is, structure A connects to structures B and C as it should, creating the right overall pattern of connectivity. However, precise neuron-to-neuron details still need to be tuned. Synaptogenesis is the formation of new synapses when neurons reach a target. That target will have what are called adhesion molecules that attract axon tips, initiating synapse formation. This process starts early in development, some five weeks after conception, but continues well after birth. Synaptogenesis seems to end somewhere between 5 and 7 years old. However, this is far from the whole story because synaptic activity continues to change. Brains are organic systems that continually adapt. Synapse formation is both activity and experience dependent; synaptogenesis can occur in later life; synaptic strengthening is an adaptive process fundamental to learning and memory. (We'll talk about this, as well as synapses downgrading, in Chapter 5.) As they develop, brains overconnect, and maintaining too many synapses (which all use energy) is neither effective nor efficient, so they are reduced or pruned back (a process called synaptic pruning). Pruning is at its maximum during adolescence, though it can continue after that. The normal effect of synaptic pruning is beneficial, though too much or too little are thought to be risk factors in conditions such as schizophrenia.

What can influence brain development, physically and psychologically? Or you could ask, What can't? There is little doubt that a portion of cognitive ability is inherited. Variability in the genome caused by such things as single nucleotide variants, SNPs and copy number variants can have an effect on cognition, though the estimated effects are usually tiny, even when consolidated together (as a polygenic measure). Still, there does appear to be a general effect of the coding genome on cognitive ability. A recent study concluded that "a large proportion (56%) of the coding genome covering all molecular functions influences cognitive abilities. One may therefore view the genetic contribution to cognitive difference as an emergent property of the entire genome not restricted to a limited number of biological pathways" [4]. That is, the proteins of which the brain is built and how they are assembled only have an overarching influence on psychological functions. On top of that is the impact of the noncoding genome and epigenetic effects and, beyond DNA, there are many other factors that have a profound effect on psychological functions.

We can focus first on infancy and childhood. We know that the genome affects the way in which a brain wires up – how axons are guided, how they find target sites and how synaptogenesis occurs. But there are two additional things to know. First, a critical period is a window in time during which certain events simply must happen. For example, in vision, depth perception has a critical period that begins at around 2 months and is at maximum around 4½ months; the critical period for spectral sensitivity (detecting light of different wavelengths) ends at around 6 months. Language acquisition has been the subject of serial arguments about the extent and nature of critical periods. Those aside, what's clear is a correlation between the age at which language is acquired and adult linguistic performance: early acquisition predicts better adult performance, implying to at least some degree the importance of timing.

Critical periods have start points and end points. They are windows of opportunity that open and then close, dependent on many factors such as axon growth, glial formation, synaptogenesis and myelination. The development of inhibitory synaptic connections may also be important. Critical periods can be harsh in that if the right conditions are not met within the right timeframe, particular functions will never develop. Sensitive periods are more forgiving, with the possibility of recovery if the optimal time is missed. It's also important to appreciate two different types of adaptive mechanisms. Biopsychologists talk about activity-dependent effects. What they mean is that neural activity itself will drive organizational changes, whether in (for example) synaptogenesis or myelination – more activity means more synapses, more myelin. Experience-dependent effects are different, reflecting the fact that an individual person's experiences can work to change brain structure and function. (This isn't just a human thing: it was found in the 1980s that an enriched environment for lab rats would improve the development of their brains, the hippocampus in particular.) What sort of experiences are we talking about?

First, the development of infants is influenced by what's around them. We tend to think of learning as a very structured thing (see the section on learning to come in Chapter 5), and it can be. But organized instruction isn't everything. Artificial intelligence engineers talk about unsupervised learning: the learning that a machine does for itself without programming or instruction. Much of what an infant does is unsupervised, even down to very basic things like the patterns of walking. Infants imitate and mimic. There are neurons in the brain called mirror neurons, present throughout the motor system (and possibly beyond) that are responsive when something is done, as well as when it's seen being done by another person – both sensory and motor properties in one neuron. Mirror neurons have an obvious capability to support imitation, which is helpful in learning about actions – how to walk, how to talk – and in acquiring knowledge about people, places and things. We learn about bodily states as well as the external world. We'll illustrate this by looking at an example of it going wrong. The following lengthy quote comes from psychotherapist Hilde Bruch's classic book, 'Eating Disorders' [5].

> I should like to cite the developmental history of an extremely fat and indolent boy, Saul, who at age 14 weighed nearly 300 lbs, at a height of 5'5". [300 lbs is around 136 kg.] According to the first information he had "always" been an insatiable eater; then it was learned that his desire for food had been without limits only since he was 10 months old. ... The boy weighed only 5 lbs at birth, was a difficult feeder and ... at two and a half months [he] weighed 7.5 lbs, and still needed to be coaxed while eating. When he was 3 or 4 months old the mother developed a backache from which she had suffered recurrently. ... The backache made it impossible for her to lean over the crib or to lift the baby. When he was able to sit in a high chair the mother still could not lift him so he sat for long stretches and would become restless and cry. The mother discovered that she could keep him quiet by sticking a cookie into his mouth. This would not keep him quiet for very long and the number of cookies increased. Saul's weight was normal at 8 months, but he had become decidedly chubby by 10 months. The mother explained that no amount of food would keep him quiet, and as her rate of feeding increased so did his weight. By 2 years of age he weighed 65 lbs, and was taken to a renowned medical center. He was placed on a 500 calorie diet and lost some weight. The family felt Saul was becoming too weak and the caloric intake was increased. The grandfather, as well as the father, made light of the mother's concern about the rapidly increasing weight, and as Saul grew older he would visit his grandfather, who would cook for him all day long (page 61 onwards).

In Hilde Bruch's words, "this history gives evidence of a grotesquely inappropriate learning experience". There was never any systematic relationship between food and Saul's bodily state. Food was a panacea for any condition, for any purpose – no amount of food would keep him quiet – at any time of the day or night – and in hopelessly unregulated amounts. Poor Saul learned all the wrong things about food, the state of his body and his emotions. This is extreme but is nevertheless anchored on what should normally happen: we learn relationships between our state, our needs and the language that describes them.

Nutrition and exercise both affect how body processes work. Getting the right balance of nutrients – proteins, fats, carbohydrates and vitamins, as well as, of course, water – all contribute to brain and body development. There is a strong link between physical exercise, diet and development. It's worth noting here something more about obesity, not as extreme as Hilde Bruch's case study but actually more concerning because of its prevalence. There are genes that influence the accumulation of adipose tissue, and there are physiological mechanisms in play (including brown adipose tissue, which we saw in Chapter 2). But a bad diet, in composition and/or volume, accompanied by a lack of physical exercise are major drivers of obesity, and there are consequences for brain development and psychological function. As a recent report from the European Childhood Obesity Group noted,

> The consequences of childhood obesity are dramatic, leading to overweight/obesity later in life, future morbidity and mortality, as well as metabolic, functional, psychosocial or quality of life impairments, among others. While most of the time unconsidered, child and adolescent obesity has been also associated with impaired brain health, structure and function, that can definitely affect individuals' day-to-day social interaction and integration, and even potential occupational success later in life. [6]

Patterns of feeding, nutrition and exercise are easy to disregard in development with trite and factually incorrect ideas about temporary 'puppy fat'. It's harder to disregard the fact that ingestion of toxins affects brain development. Lead is a notorious problem, coming from lead pipes, lead paint and the exhaust fumes of cars using petrol with lead added. All of these can be eliminated, and while in many countries, the risks have been significantly diminished, lead poisoning remains a global challenge. Other airborne pollutants from industrial processes, agricultural pesticides or transport emissions can influence development. Their effects will vary by location, and their scale remains a matter of debate, but the fact that pollutants have consequences for development is clear. We have already mentioned fetal alcohol syndrome, and there are well-documented cases of newborns who receive crack cocaine or heroin through the placenta in the womb and experience unpleasant withdrawal symptoms at birth as soon as the blood supply from the mother is ended. And as if all this malign chemistry wasn't enough, there are multiple studies looking at the negative effects of low socioeconomic status and inadequate education (of both children and their parents) on physical and psychological development.

When childhood ends, puberty starts with a new set of developmental hurdles. The onset of puberty is earlier in females than in males: between 10 and 17 years old for females compared to 12 to 18 for males. (The wide range tells you something about how difficult it is to define puberty by time alone.) Adolescence is another hard-to-define stage. Hormonal activity, sexual maturation, changes in body shape and growth in general are all part of it. We talked about hormones in Chapter 2, when we discussed the endocrine system and sexual reproduction. Hormones have effects over many years. During prenatal development, estrogen and testosterone guide the sexual differentiation of the brain. Brain-derived neurotrophic factor, a substance found in multiple places around the body and which can operate as a hormone, influences synaptogenesis and promotes the growth and survival of neurons, leading to the establishment of neural circuits. In adolescence, sex hormones drive physical changes, creating differentiated sexual characteristics, as well as influencing synaptic pruning in the brain. Mood, emotion, aggression and stress are all susceptible to changing levels of hormones in males and females. In addition, over time, female sex hormone activity fluctuates with the menstrual cycle and changes during menopause.

"Give me a child until he is seven and I will show you the man" is a claim that comes from Aristotle, though it's often attributed to St. Ignatius Loyola, founder of the Jesuits, as well as being adopted by others more scientifically minded. Is it really true? Experiences in childhood, whether physical events in the brain and body, sensory, social or psychological have an effect on cognitive and emotional development. But development is a long and idiosyncratic journey, not guaranteed to fit a precise timeline. The multiplicity of factors that impact physical and psychological development during childhood and adolescence is important. No single one of them, acting alone, is likely to be crucial, but put them all together, each having independent as well as interactive effects, and things start to happen. The polyfactorial nature of influences on brain, behavior and psychological life guarantees that if you take just one factor, you'll inevitably find someone who claims, "I did that, and it never did me any harm", or "I overcame disadvantage, so everyone else can". Great: but the shaping of development to make us as individual persons is idiosyncratic to the max, multiple factors interacting. What makes you, you, is yours alone.

LIFELONG DEVELOPMENT AND AGING

Brain development is not limited to childhood and adolescence. The brain continues to develop and adapt throughout life through processes of synaptic strengthening (which we'll talk about in Chapter 5) and even adult neurogenesis. There has been discussion recently about when the brain reaches its final adult form. The claim that this doesn't happen until 25 years old is being used in some places to argue that, for example, criminal sentencing should be more lenient to the under-25s because their brains aren't yet fully developed. But the proposition that the brain doesn't reach its final form until 25 is something to be very cautious about, not to say outright skeptical. A key driver of this claim comes from structural scans of brains [7]. (If you look into this paper, you'll see that in Figure 3, there is a time line of brain development. The difference in measurements between late childhood [6 years] and young adulthood [25 years] is negligible.) Measurements include things like the volume of cortical white and gray matter, cortical thickness, total surface area, the volume of the ventricles and the relative size of the tissue under the cortex. But there are two caveats: first, the degree of interindividual variability in these measures is substantial, and second, what is not measured by the scans? Fine details are missing about things that have an obvious effect on brain function, such as the structural details of synaptic organization, myelination and transmission speeds. So here's a question: are you older than 25? If so, did you wake up on that birthday feeling smarter, more responsible, more adult? No, neither did we.

Biopsychology cuts across the lifespan. Adopting sensible behaviors regarding nutrition, exercise, mental activity and lifestyle all contribute to health. The first-century Roman poet Juvenal nailed it with *orandum est ut sit mens sana in corpore sano*, a liberal translation of which is that we should pray for a sound mind in a sound body. It seems like stating the really obvious, but sensible diets, mental and physical exercise and an avoidance of substance abuse are needed to maintain health and promote longevity. Biopsychology can help when things go wrong. Treatments and therapies for behavioral, psychological, psychiatric and neurological disorders obviously feature. Counseling, behavior modification, rehabilitation programs, pharmaceutical and medical treatments – including contemporary neuroprosthetics, brain-machine interface devices – all have roots in biopsychology. We'll look at all these in some detail in Chapter 7.

Aging is inevitable. All cells gradually lose the ability to maintain themselves – biologists talk about this form of aging as cell senescence. Various forms of death overcome cells. The most common distinction is between apoptosis and necrosis. Apoptosis is a form of genetically programmed cell death in which cells are literally dismantled, while necrosis is accidental, the result of something going wrong. Unlike apoptosis, which is regulated and does not involve inflammation, necrosis is usually inflammatory. The cell senescence of normal aging can be thought of as routine wear-and-tear, but disease can accelerate damage to organs through necrotic processes, and even therapies can have negative effects. Radiotherapy, for example, or organ transplants, can compromise the immune system and accelerate cell senescence. Brains atrophy as they get older, with loss of cells and white matter. The numbers of synapses are reduced, and neurotransmitter production slows. Tissue shrinkage is normal in aging, though where it occurs varies. Reduction in the volume of the temporal lobe and the hippocampus is likely to lead to problems with memory, but shrinkage in the frontal lobes gives rise to another common problem of old age: loss of inhibition (when grandparents start to tell it like it is, no holding back).

The other thing that can happen as brains age is neurodegeneration, the loss of neurons triggered by issues in the genome or via environmental triggers epigenetically. There are a variety of neurodegenerative disorders that can affect sites from the brainstem to the cortex and everything in between, as well as demyelinating disorders that can spread widely. Most of these disorders involve the activity of several genes, the exception being Huntington's chorea, which has a regrettably straightforward inheritance. It is caused by a mutation in a gene on chromosome 4 that codes for a protein called huntingtin. If one parent has the gene, there is a 50:50 chance that their offspring will have it too. In Alzheimer's disease, having the gene *APOE-e4* is a major risk, though there are many other identified genes associated with late-onset Alzheimer's. Several genes are risk factors in Parkinson's disease. The gene *SNCA* makes a protein called alpha-synuclein, which accumulates in clumps called Lewy bodies, while another gene, *PARK2*, makes a protein called parkin, which is defective in this disease. We'll discuss these neurological disorders more in Chapter 7, but there is one thing worth mentioning here because it links biology and psychology.

It's about Alzheimer's disease. In this condition, the brain develops plaques dense in amyloid protein and neurofibrillary tangles that look like little knotted ropes. Plaques and tangles develop over time, and the degree to which they're present is used to identify the stage to which the disease has progressed. However, it has become increasingly clear that a condition called asymptomatic Alzheimer's exists in which the classic physiological hallmarks are present in the brain – sometimes to a very high degree normally predictive of late-stage Alzheimer's – but without significant cognitive impairment. How is this possible? On the one hand, it might show that there is a yet unidentified cellular factor involved in the disease that is present in those who have it but not in asymptomatic cases. On the other hand, it might be evidence of neuroplasticity, the ability of neural tissue to adapt and change in response to events. We've seen this in activity and experience-dependent changes, and we'll see more of it in the next chapter when we talk about synaptic plasticity and memory. And on a third hand, maybe life events are predictors of the ability to overcome Alzheimer's neurodegeneration. People have looked at factors such as sex, socioeconomic status and education as predictors. All seem to have an effect, though each one is rather small. A more recent claim is that what's called eudaemonia well-being is a stronger predictor than any of the others. What is it? Hedonia is a feeling of transient pleasure — a good meal, your favorite team winning. Eudaemonia, on the other hand, is a deeper feeling of personal autonomy, control and satisfaction with life [8]. Whatever the reason, the pathology of Alzheimer's does not of necessity predict having the symptomatology. The same may be true of other disorders. Pathologists have always reported the presence of Lewy bodies in the brains of people who did not show Parkinson's disease, the assumption being that these individuals were presymptomatic.

Aging is a complex process in which cells die, genes misbehave and environmental events make their impact. How this constellation of events plays out for any one individual is likely to be more idiosyncratic than one might have expected, biology and psychology bouncing off each other in ways that aren't easy to predict.

SLEEP

We're going to end this chapter on development with sleep. The reason we're discussing sleep as part of development is that sleep affects everything. It's important both during early development and across the lifespan; it's involved in cognition; it affects emotional and social life and is associated with all sorts of different psychopathologies and neurological conditions. It has a ubiquitous effect on everything we do. We need to talk about it, and now's the best time.

The structure of sleep. Sleep isn't all one thing. The two major divisions are slow wave sleep and rapid eye movement sleep (REM), defined by various measures including brain waves, eye movements and muscle tone. The major sleep components are presented in Table 4.1.

What's striking is how similar REM sleep and the waking state are. The resemblance is in the desynchronized patterns of activity, with the cortex showing similar electrical activity in both states. The difference is in the presence of unique waveforms (such as the PGO spikes) and in the suppression of muscle activity. It's as if the brain was awake, but the body switched off. Dreaming happens during REM sleep – REM sleep is often referred to as D sleep, but D stands for desynchronized, not dreaming – and you will be aware

Table 4.1 The structure of sleep

Minutes	Condition	Sleep stage	Features
	Awake		Alpha and beta waves
0	Drowsy	SWS stage 1	Theta waves
+10	Asleep	SWS stage 2	Theta waves, sleep spindles, K complexes
+15		SWS stage 3	20%–50% of EEG is delta wave
+20		SWS stage 4	>50% of EEG is delta waves
+45		REM sleep	• Theta and beta waves (like the awake state) • EEG desynchronized (unlike synchronized SWS) • High EOG activity • Low EMG activity: muscle tone reduced • Increased brain blood flow and O_2 consumption • PGO spikes present • Possible genital activity • In humans, typically 4–5 REM bouts per 8h sleep

Abbreviations and explanations

EEG	Recording brain electrical activity using scalp electrodes Electroencephalogram – the EEG record; electroencephalograph – the machine; electroencephalography – the process
EMG	Electromyogram, a recording of muscle tone
EOG	Electrooculogram, a measure of eye movements
REM	Rapid eye movement
SWS	Slow wave sleep
Alpha waves	Regular, medium frequency (8–12 Hz). Synchronized: at rest, eyes shut
Beta waves	Irregular, low amplitude, high frequency (13–30 Hz) Desynchronized: alert, attentive, prevalent during mental activity
Delta waves	High amplitude, low frequency (<3.5 Hz)
Theta waves	Medium frequency (3.5–7.5 Hz)
Sleep spindles	Bursts of 12–14 Hz activity, 2–5 times /min
K complexes	Sudden burst of activity, approximately 1 per min, often associated with stimuli (such as noise) and only present in SWS stage 2
PGO spikes	Pons-geniculate-occipital activity: characteristic of REM sleep

of people, places and things that you know, albeit in a disorganized way.

Brain mechanisms of sleep. There is no single definitive location or chemical that controls sleep. Rather, it is distributed property. Deep brain structures have particular importance. In Chapter 1, we talked about the ascending reticular activating system. Activity in this is closely associated with behavioral state control – that is, switching between the three states in which we exist: awake, slow wave sleep and REM sleep. As we noted before, the ascending reticular activating system has neurons in several places through the brainstem. The long axons of these neurons go everywhere and use monoamine neurotransmitters like serotonin, noradrenaline and dopamine. One target is the thalamus. Thalamic neurons exist in two states: single spiking and burst firing. Single spiking – spontaneous single-action potentials – is associated with information transfer to the cortex, wakefulness and REM sleep. In contrast, burst firing – bursts of action potentials at 7–14 Hz – are rhythmic, encouraging cortical synchronization, sleep spindles and slow wave sleep. The ascending reticular activating system has a large degree of control over the thalamus and, consequently, the cortex, control that is pivotal in behavioral state control. There's more. In the lateral hypothalamus are neurons containing a neurotransmitter called hypocretin/orexin. (Hypocretin and orexin are the same thing, a peptide. Two labs discovered the peptide almost simultaneously. One lab called it orexin because it stimulated feeding; the other called it hypocretin because it's structurally related to another peptide, secretin [9].) The neurons containing hypocretin/orexin have extensive projections into the brainstem, midbrain and cortex, as well as neurons in the ventrolateral preoptic area (at the front end of the hypothalamus). These neurons project to the tuberomammillary nucleus (at the back of the hypothalamus). This contains histamine neurons that project to the cortex and to the ascending reticular activating system. The interplay between these systems is a flip-flop one: tuberomammillary neurons on, ascending reticular activating system and hypocretin/orexin systems off. It's complicated, but there's a simple take-home message: multiple brain systems interact to regulate sleep.

Sleep disorders. Sleep is remarkably idiosyncratic. The world record for sleep deprivation was set in December 1963 by a 17-year-old called Randy Gardner, who was under supervision in the laboratory of William Dement, a noted sleep researcher: 11 days, 24 minutes – unbelievable. Afterward, Randy went through REM rebound. Typically, after sleep deprivation (on a normal scale), the missed hours are not recovered, but there is an increase in REM sleep in the nights following sleep loss. An average normal night's sleep for an adult human is seven or eight hours, though for some people, it's as little as three hours, and for others, it goes into double digits. Infants sleep more than adults, and the elderly spend more time in bed, though the quality of sleep declines – it becomes more fragmented – and there is likely to be more daytime napping [10]. So, aside from the idiosyncrasy of sleep time and the effects of deprivation and aging, what are the main sleep disorders?

- **Insomnia** is the inability to sleep. Onset insomnia is the inability to fall asleep; maintenance insomnia involves frequent waking; termination insomnia is early waking with an inability to resume; drug dependency insomnia is induced by substance abuse – paradoxically, often sleeping pills.
- **Sleep apnea** involves difficulties in breathing, with an abnormal accumulation of carbon dioxide leading to sudden waking. Central apnea is produced by changes in the brain (including drug-induced ones); obstructive apnea involves relaxation of throat muscles, which obstructs airflow. It's often associated with snoring, which is itself a considerable disruptor of sleep (including other people's sleep).
- **Narcolepsy** is a REM disorder composed of sleep attack (very sudden onset of sleep), cataplexy (loss of muscle tone but no loss of consciousness), sleep paralysis (an inability to move immediately before or after sleep) and hypnagogic and hypnopompic hallucinations. These are vivid hallucinations occurring immediately before (hypnagogic) or after sleep (hypnopompic). Narcolepsy is particularly associated with a deficit of the neurotransmitter hypocretin/orexin. Dogs are unusually prone to it. A recessive gene is linked to narcolepsy and is present most commonly in Doberman pinschers, though other breeds have it too.
- **REM sleep behavior disorder** happens when normal muscle activity is maintained during REM sleep rather than suppressed

as it should be. It leads to acting out of dreams (such as kicking out when one scores a dream goal).
- **Three disorders associated with slow wave sleep**: nocturnal enuresis – bedwetting, somnambulism – sleepwalking, pavor nocturnis – night terrors.
- **Jet lag** is a circadian rhythm problem. We talked about the brain mechanisms of circadian rhythms in Chapter 3, but we need to add something. Melatonin is a hormone released into the brain from the pineal gland (behind the thalamus, on top of the brainstem – Descartes thought it the seat of the soul, but it isn't). The pineal gland is responsive to the light-dark cycle and regulated by the suprachiasmatic nucleus (the key circadian controller). Shifting across time zones disrupts circadian rhythms, leading to a sleep/wake cycle out of kilter with the day/night one. Taking melatonin supplements may (we stress, may) help adjust to new time zones, diminishing jet lag.

It's also worth noting that disturbed sleep is a common feature of many neurological and psychiatric disorders. However, in no case is it truly diagnostic. A bad night's sleep is not the precursor to illness.

The function of sleep. If you look around all vertebrate animals (and often beyond), you'll find that sleep is universal. Species evolve, adapting and changing, but whatever happens, sleep persists. Different species sleep for different amounts of time, in part depending on their rank in the food chain, and sleep patterns are different. To prevent drowning, marine mammals can sleep one hemisphere at a time, tending to swim in circles with the awake hemisphere on the alert. Birds that travel long distances airborne also sleep one hemisphere at a time. No animal can do without sleep. Our productivity, in school, in the lab, in whatever occupation we have, is influenced by sleep habits as much as they are by diet and lifestyle. Poor quality sleep impairs reaction times, judgment, attention and memory and has a general negative effect on health. Regular hours, quiet, low light and leaving time after a meal to digest all help a good night's sleep.

But why do we sleep? Why does the average person spend literally a third of their life not awake? Memory consolidation? Protein synthesis and brain restoration? These are the usual suspects, but in truth, sleep is a mystery, an idiosyncratic, essential mystery.

CHAPTER 4: SUMMARY

- *A gene is a sequence within a DNA molecule; genomics is the study of DNA (and a genome is all the genes in a cell, an organ or a body); genetics is the study of inheritance; a genetic code is for the manufacture of proteins; chromosomes (two Xs in biological females, X and Y in biological males) are in cell nuclei and have all the genetic material.*
- *The coding genome is the part that codes for proteins; the noncoding genome (much the bigger part) is concerned with epigenetics – how genes are read, how they can be turned on (gene expression) or off (gene silencing). Events in the body or in the environment have epigenetic effects.*
- *Brain development works from a neural plate to a neural tube that will form the brain and spinal cord. Neurons and glia come from stem cells; myelination is the deposition of myelin to electrically insulate axons; synaptogenesis is the formation of contacts between neurons, a process that is cut back in adolescence (through synaptic pruning).*
- *Psychology and biology interact: genes have significant effects on brain development, but there is also activity and experience-dependent development (brain activity driving change; what a person experiences in life changing their brain).*
- *Some development is lifelong; aging inevitably involves the natural loss of brain cells and may also involve neurodegeneration, leading to behavioral and psychological effects.*
- *Sleep is important for health throughout life. It is not a unitary event but divides into distinct phases (REM and slow wave sleep). There are known brain mechanisms; the time for which a person sleeps is idiosyncratic, and disorders of sleep can appear on their own or be part of other neurological and psychiatric disorders.*

COGNITION

Scientists often talk about "the provisional nature of knowledge". They mean that what we know right now might not be how we see things tomorrow because scientific knowledge advances. The biggest example we can think of is truly astronomical: in the sixteenth century, science shifted from Earth being the center of everything to a position in which the sun was the agreed center of our solar system. The change was driven by the genius of Nikolaus Copernicus (and refined by Johannes Kepler), who grasped that increasingly detailed observations of the transits of stars and planets could not be explained in a geocentric universe – Earth at the center – but were better accommodated by a heliocentric one – sun at the center. Developments in scientific knowledge are not usually this radical and usually happen gradually or in small increments, with slow accumulation of new data. But when so much new data (perhaps driven by new technologies and better measurements) can't be explained, a revolution must happen and a new paradigm established. What's a paradigm? A conceptual framework underlying the theories and practices of a discipline at a particular time, an accepted overarching view. The idea was introduced by Thomas S. Kuhn in his landmark work 'The Structure of Scientific Revolutions' (1962). It has stood the test of time, and every scientist should read it. Paradigms structure how we create and develop theories, how we understand information and how we educate new generations of researchers.

In biology, Theodore Schwann's discovery that animals are made of cells, or Charles Darwin's ideas about evolution, changed everything and established new paradigms for understanding bodies and the natural world. Grasping the double helix structure of DNA was paradigm shifting, creating molecular biology. Genetics (the study of heredity) and genomics (the study of DNA) have been transformed by the realization that the genome is not fixed but is more adaptable than expected. The coding genome makes proteins, and a non-coding genome engages in the repair and modification of proteins through epigenetics. (We discussed these in Chapter 4.) Because scientific knowledge and explanation sometimes change dramatically, it means that a basics book like this can be caught on the horns of a dilemma. There is established and slowly growing information that students and practitioners ought to know, set within current theories and paradigms. Then there is direction of travel: where are we headed? Are big changes imminent? You get the drift. For the biopsychology of cognition, we'll outline the basics, but we think change is coming.

You'll recall that in the Introduction we talked about the development of psychology as a discipline and how cognitive science regained the study of internal, mental processes after a period in which behaviorism had been dominant. Psychological paradigms shifted and computer science was a big driver. Cognition has come to refer to the mental processes involved in acquiring, processing and using information, which sounds very computer-esque. The term cognition includes a wide range of activities, such as perception, attention, memory and language.

Essentially, cognition refers to how we think, understand, and interact with the world around (and inside) us. There's an interesting point to note before we go any further. Judging from their behavior (which objectively is all we have to go on), other animals share cognitive abilities with us. They might differ in degree, but we can be confident that all sorts of animals, vertebrate and invertebrate alike, perceive the world, pay attention, learn and remember, make decisions, communicate and so on. Can we describe these as mental events in animals? Hard to say because the internal lives of others are at best opaque and more usually totally hidden. They must be dealing with sensory input, but does 'information processing' mean the same as 'mental events', or are these separable, totally or in some degree? Can we make assumptions about the ways in which humans and animals 'do' cognition – must they be the same, one size fits all? As a problematic instance, we recognize the intelligence of octopuses from their behavior but their nervous systems are structured in a radically different way to that of vertebrates. And of course this isn't just about animals. Artificial intelligence raises the same kinds of questions, and if you want to go really wild, would extraterrestrial aliens have the same cognitive lives as us? These are problems of some considerable scientific and philosophical depth. All we're doing here is alerting you to the issue: what we call human cognition need not necessarily be the same thing that happens across every species.

We're going to approach cognition by starting with memory. It might seem odd to make this our first choice, but it's important because memory informs everything else that's included under the word 'cognition'.

MEMORY

Memories help us to make sense of the world and provide us with a sense of continuity and coherence in our lives. While our memories may not be perfect, they are an important part of who we are and play a critical role in shaping our personal narratives and our understanding of ourselves and the world around us. Human memory is an amazing thing. Based on data about the connectivity of synapses, it's been estimated that the human brain has the potential to hold up to a petabyte of data, which is 1,000 terabytes or 1,000,000 gigabytes, equivalent to the whole of the World Wide Web [11].

Such statistics make for great press releases, but it's healthy not to get too carried away for at least three reasons. First, synaptic connections aren't all about remembering facts but have to do with actions as well. How we walk, how we talk, how we manage them both simultaneously – these and other behaviors are learned and remembered from infancy onward. Second, memories are not recorded as a video camera records a scene but are reconstructed each time we recall them, and the process of reconstruction can be influenced by a variety of factors, such as our mood, beliefs, and expectations. What we're storing isn't necessarily as fixed as we might think. And third, human working memory, the system responsible for holding and manipulating information in mind over the very short term – shopping lists, bits-and-pieces you need to do right now – isn't much different from that of a lab rat. The average person can hold about seven items in their working memory at any given time.

We obviously have memories of people, places and things. Some things we remember well, others slip away quite quickly. Sleep plays a critical role in memory consolidation and emotionally charged, or especially significant events are more likely to be remembered in comparison to neutral ones. A simple test: try to recall what you had for breakfast yesterday: likely you can. Breakfast on a random day a few months back? Likely you can't. But on your last birthday, or the first day of a vacation or some other special day a while back? That might have stuck despite the passage of time. Nevertheless, however well consolidated, memories are not always accurate. They can be distorted or change over time and can be influenced by our current state. For example, we may remember an event as being more positive or negative than it actually was, depending on our emotional state at the time of recall. We may also fill in gaps in our memories with information that is not actually part of the original event. All of this fluidity is a key part of what makes memory so fascinating to study.

HOW WE CLASSIFY MEMORIES

Psychologists classify memory into different types, and neuroscientists look to associate them with different parts of the brain – the so-called multiple memory systems. Long-term and short-term memory are basic concepts, but what we're looking at here has more to do with the nature of what's being remembered. Categories can change over time, but these are some of the most important.

- Declarative memory refers to our ability consciously to recall information about facts, events and concepts. Declarative memory can be further divided. (i) Episodic memory is our ability to remember specific events and experiences, such as a vacation, first date or Champions League final; the hippocampus is a focus of research here. (ii) Semantic memory is the ability to remember general knowledge and concepts, such as the capital of a country or the meaning of a word. (iii) Autobiographical memory refers to our ability to remember and mentally revisit events and experiences from our own lives. It includes elements of both episodic and semantic memory.

- Procedural memory relates to our ability to learn and remember how to perform a task or a skill, like riding a bike or playing a musical instrument. It's often overlooked because it can be too easy to think of memory in terms of school, learning and remembering facts about things – what, where, why, when, who. But learning 'how' is a critical part of memory, from the simple act of walking, through manipulating food in order to eat, to specifically human activities like using cutlery properly, piano playing or doing neurosurgery. Sometimes procedural memory is thought of as implicit memory (we just do it) in contrast to declarative, which is explicit memory (and needs recall, aka retrieval). The brain's motor systems are important for procedural memory.
- Spatial memory refers to our memory of space and place. It's necessary for navigation through the world. Spatial memory operates on both an egocentric basis and an allocentric one. Egocentric is our knowledge of where things are in relation to us and our current position; allocentric is to do with where things are in relation to each other. For example, the authors know where the university library and sports center are in relation to our departments, and we know how to get to any of them from where we are right now. There has been a vast amount of research on spatial memory and the hippocampus in the last decades.
- Sensory memory refers to our ability to briefly retain information from our senses, such as what we see, hear, or feel. This appears to be an intrinsic part of any sensory system: does it raise a question about the separation of sensation and cognition? We'll see.
- Working memory is our ability to hold and manipulate information in our minds while we perform a task, such as following directions or remembering a phone number while we dial (as if people do this anymore). The prefrontal cortex is thought to be important for this.

THE BIOLOGICAL BASIS OF MEMORY

The assumption made about human memory is that it follows a three-step sequence: encoding, storage and retrieval. Encoding and storage are both accepted as being done by neurons, with emphasis in particular on the connections between them. When neurons communicate with each other, they can establish small-scale ensembles and larger-scale neural networks (some of which will be fixed, others of which might drift over time) that are responsible for encoding and storing information. Central to this is synaptic plasticity, the strengthening of connections between neurons. Well before the mechanisms were revealed, the Canadian psychologist Donald Olding Hebb proposed the idea of synaptic strengthening – "cells that fire together, wire together". The realization of Hebb's idea – the so-called Hebbian synapse – began with the discovery by Tim Bliss and Terje Lømo that a train of electrical impulses to neurons altered their responsiveness on a long-term basis, for hours, days or (as far as can be known) permanently. This was called long-term potentiation (LTP). The reverse also exists: long-term depression (LTD), a persistent decrease in the reactivity of neurons. In essence, LTP means that when two neurons are repeatedly activated at the same time, the strength of the connection between them increases. Contrastingly, LTD weakens the connections between neurons that are not used frequently. This is thought to be important because it allows the brain to trim away unnecessary or redundant information.

Both LTP and LTD depend on multiple neurotransmitter actions. The amino acid transmitter glutamate is a central player, acting at different types of receptors during the formation of LTP, but others such as dopamine, noradrenaline, serotonin and acetylcholine also play critical roles in modulating synaptic strengthening. Synaptic plasticity is very complex and involves not just neurotransmitters and their receptors but changes in gene expression as well. Epigenetic modifications are changes in gene expression without changes to the underlying DNA sequence (discussed in Chapter 4), and these modifications play a role in memory formation and maintenance. For example, the expression of certain genes in the hippocampus can lead to improvements in memory. In addition, glial cells have a role in supporting synaptic activity associated with memory processes, with many recent studies finding that astroglia release molecules at synapses to enhance memory formation. It is unambiguously the case that neurons are central to memory, but they never act alone.

LTP was first observed in the hippocampus, and much of the early work on it focused heavily on this part of the brain. Richard Morris and his colleagues, for example, were able to show that blockade of LTP formation in the rat hippocampus abolished the ability to form new spatial memories. However, it has become abundantly clear that LTP and LTD are universal and not confined to just one brain structure. Let's be clear: memory is not confined to a single location or "center" in the brain. It's likely that different brain structures are involved in different sorts of memories. The hippocampus still looms large in memory research. The 2014 Nobel Prize in Physiology or Medicine was awarded to John O'Keefe, May-Britt Moser and Edvard Moser for their discoveries of hippocampal cells that constitute a positioning system in the brain: that is, spatial memory. Research has shown that the amygdala is involved in emotional memories, the prefrontal cortex in working memory and the cerebellum in the formation and storage of procedural memories. But a word of caution: while the principle of different functions for different parts of the brain holds good, one does need to be careful of overidentifying specific functions with specific places. The brain is a prodigiously interconnected organ and no part of it is an island, entire of itself (to slightly misquote John Donne). Everything that goes on in a particular structure depends on inputs from other parts of the brain.

We can confidently assert that repeated activation of groups of neurons (whether a small ensemble or a large network) leads to a strengthening of connections between the neurons that make up the entity. This results in the formation of what scientists refer to as an engram or memory trace, which represents the information that was learned or experienced. Engrams remain largely hypothetical concepts because, while the brain does code and represent information, it is not yet possible (if it ever will be) to tag the memory of a very specific thing to very specific neurons. There are three stages in creation of an engram: first, input of information into the network; second, association of the new information with existing information in the network, achieved through the strengthening of synaptic connections between neurons that are activated together; and third, output, the retrieval of the memory. Biologically, this seems entirely reasonable, paralleling the kinds of processes that might occur in a computer. Interference with this neural machinery is more than likely to cause memory impairments of one sort or another, though 'impairment' has to be more than the loss of a single neuron. What makes ensembles and networks robust is their connectivity, not the operation of one single neuron. However, after input and encoding there is an issue that troubles some philosophers: that there is an explanatory gap between physical and psychological properties. How do physical facts about a person's brain determine that they are having any psychological experiences at all? We can talk with some confidence about neuronal activity providing biological substrates for memory, but can we completely explain in biological terminology what happens when we retrieve a memory and bring it to mind? This is the explanatory gap, the apparent disjunction between matter and mental events. However, while this troubles some, other philosophers dismiss such concerns as remnants of a false mind/body dualism (that is, mind and brain are separate entities). We don't have a categorical answer to this (any more than anyone else does) but the question is important and worth your while following up for yourself.

HOW WE GET THINGS INTO MEMORY – LEARNING

It's not possible to separate learning from memory. It's usual in everyday biopsychology talk to speak of 'learning and memory' in the same breath. There's even an academic journal called 'Learning & Memory'. Just as it's possible to describe the brain machinery of memory, we can talk about different mechanisms of learning as developed by behaviorists using lab animals as subjects. Two basic forms of learning are Pavlovian and instrumental. In the introduction to this book we met their most influential developers: I. P. Pavlov and B. F. Skinner. Pavlovian conditioning in its initial setting involved ringing a bell at the time a dog was waiting to be fed. The bell predicted food, and so in due course, the sound of the bell caused the dog to salivate. Theorists speak of the food being an unconditioned stimulus and the bell as a conditioned stimulus. Instrumental learning (aka operant learning or operant conditioning) was developed by Skinner and others and differed in that an animal was required to commit an action by operating (hence operant conditioning) an instrument (hence instrumental conditioning). Skinner developed an operant chamber (aka Skinner box) as a totally controlled environment in which the actions that could be emitted by the subject were restricted and the stimuli it received controlled. Simple experiment: press a lever, get a food treat as a reward. This could be elaborated by adding extra levers, lights going on or off and tones being sounded, all with the aim of letting the subject learn what to do under specific experimental contingencies. Central to Skinner's work were four basic pillars: schedules of reinforcement. They are as follows:

- Positive reinforcement: for example, press a lever and obtain a reward. The delivery of the reward reinforces the act that produced it, making another lever press more likely. Note that reward and reinforcement are related but not the same: a reward reinforces the action that delivered it.
- Negative reinforcement: this is often misunderstood. In this kind of learning, the subject presses a lever in order to avoid something bad, so like positive reinforcement, rates of behavior need to be high – keep pressing to avoid unpleasantness.
- Punishment: this is what many assume negative reinforcement to be. Punishment is what it says: press the lever, and something bad happens. The rate of behavior therefore goes down, not up.
- Omission: rates of lever pressing go down if rewards stop being delivered, simple.

Schedules of reinforcement refer to the way in which outcomes are delivered. Just thinking about positive reinforcement and lever pressing, fixed ratio (FR) schedules have a reward delivered for a fixed ratio of actions: FR10 means one reward on every tenth press. Variable ratio (VR) schedules differ, in that VR10 would have a reward delivered on average every tenth press. In contrast to these, fixed interval and variable interval schedules use time rather than the number of actions to determine when rewards are delivered. More complex schedules also exist: progressive ratio schedules start low but increase stepwise. One press earns the first reward, but the next one needs two lever presses, then four and so on. The usefulness of this is in the study of motivation – where's the break point at which the lab rat decides that the cost isn't worth the outcome?

Theories of learning developed from basic principles like these but quickly became very much more complex. We needn't dwell on the intricacies, and we'll be honest and admit that work like this, very behaviorist in nature, has gone out of fashion in biopsychology.

Nevertheless, it's still important for a number of reasons. First, as we saw in Chapter 3, operant boxes are still used in psychopharmacological research to test learning and memory, as well as processes like attention. The boxes themselves are much more sophisticated, often now incorporating touch screen displays rather than mechanical levers. Second, behaviorist principles of learning had a significant impact on clinical practice. Behavior modification techniques are still employed in the treatment of phobias or anger management, for example. Third, behaviorist theories have informed both the neurobiology of learning and memory and machine learning. In both of these, key terms frequently used are association (associative learning – putting events and actions together, whether Pavlovian or instrumental) and reinforcement.

The principles of associative learning and reinforcement learning are very consistent with what we now know about synaptic strengthening. Events and actions become associated, learned and remembered through the changes that happen at synapses. Machine learning likewise has adopted both associationist and reinforcement learning as key constructs. It is increasingly used to make predictions, estimations and recommendations around an identifiable task (such as in sales, for example: "If you liked that thing, you might like this one too"). The machine uses past data to solve problems and deliver best-fit outcomes. It might operate using supervised learning. In developing accurate recognition of, for instance, dogs versus cats, examples are given, and the machine learns by using its errors to correct its future performance (an error reduction process that's known in neural network circles as backpropagation). In contrast, unsupervised learning goes on without explicit training using labeled examples. Likewise, reinforcement learning by networks relies on experience to produce appropriate reinforcement that will shape future performance.

Machine learning relies heavily on ideas drawn from neuroscience, but equally, neuroscience has brought ideas back. Concepts like supervised and unsupervised learning or reinforcement learning, seem to have a more human feel in comparison to the very dry theories of behaviorism. Perhaps the most interesting is unsupervised learning, given that so much of what we do in the world and what we know about it is acquired with minimal training. Children in particular just sponge stuff up.

WHY DO WE FORGET?

Forgetting is a complex process that involves multiple brain regions and neural mechanisms. It's an everyday thing, and we do it all the time. It becomes amnesia when learning and memory deficits become clinically concerning (retrograde amnesia being an inability to remember past events, anterograde amnesia being the inability to form new memories). Beyond biopsychology, other kinds of psychological theory have their own approaches to this. Psychoanalysts might treat forgetting as a mechanism for avoiding recall of a traumatic experience – for example, a deliberate, if not necessarily conscious, suppression. Freud talked about people 'suffering from reminiscences', an apt description of how memories can be painful (for a good modern account of Freud, see [12]). However, in biological psychology, we can identify several elements of forgetting.

- Interference: one major reason for forgetting is interference from other memories. When similar memories compete for neural resources – networks with overlapping elements, for example – they can interfere with each other and cause forgetting. This is especially true for memories that are encoded closely in time and space.
- Synaptic decay refers to the weakening of the synaptic connections. If a memory is not accessed or used for an extended period of time, the synaptic connections that support it can gradually weaken, leading to forgetting.
- Disruption of consolidation can occur due to factors such as sleep deprivation, stress, or brain injury. You try to consolidate a memory, but it doesn't happen – it's one reason why a notebook by the bed is useful for clinging to ideas in the middle of the night.
- Retrieval failure can occur when the cues or contexts that were associated with the memory during encoding are not present during retrieval.
- Aging is associated with a reduction in the number of synapses and alterations in neurotransmitter levels. Neurodegenerative disorders such as Alzheimer's disease can cause progressive and widespread damage to the brain, impacting memory.

SENSATION AND PERCEPTION

Memory is the bedrock of cognition. It's often said that 'vision is cognition', meaning that we have to learn and remember how to identify what it is we see. The same could be said for any other sensory system – we learn and remember about sounds, tastes and everything else. What they are, what they mean, what they feel like – all of it learned and remembered. However, this all relates to the content of our senses. What about the machinery? It's important that we talk about that but first a question: how many senses do humans have? Vision, hearing, taste, touch, smell: five senses. Five is the classical answer, and it goes right back to Aristotle, yet despite the antiquity, it's not right. These five are all exteroceptive – that is, they deal with things from the outside world and our interactions with them. What's missing are interoceptive senses, the things that come from inside, such as equilibroception, the sense of balance, and proprioception, the sense of where our bodies are in space, our position. As well as these of course, there are the sensations that come from body organs. The vagus nerve, for example (see Chapter 1), brings all sorts of sensory information about body physiology into the brain. What's more, you could, if you wanted, sub-divide the classical five – a sense of color, for example, which isn't the same as a sense of movement in the visual field. Pain could be separated from other skin sensations. Or we could group things differently: the term haptic refers to all of touch, proprioception and kinaesthesia, the sense of motion. The idea of five senses seems suddenly less viable, doesn't it? And we haven't even considered other species in which senses might be different or altered. Cats can't detect sweetness; dogs don't have the visual elements needed to see red and green, only blue and yellow; rats detect events literally in front of their noses using their long whiskers (aka vibrissae); octopuses use their tentacles for a sense that combines touch and taste. However, we're not going to deal with senses one by one but instead try to deal with them collectively, establishing some basics of sensation rather than concentrate on the intricate details of each one.

The first step in sensation is the detection of events: light in the retina activating rod and cone cells, air pressure in the ear, soluble molecules in the nose and on the tongue, mechanical pressure or changed chemistry on the skin, or events in joints, tendons, muscles or blood. Often, it will involve specialized receptors able to detect a particular event, whether the wavelength of light in the retina or sweetness on the tongue; for sound, physical distortion of the basilar membrane in the ear; and for muscles and body organs, stretch receptors. However it is that events are detected, the receptive mechanism converts them into electrical impulses that travel along nerves. This is called transduction. (See Chapter 1 for more discussion of peripheral nerves.)

Sensory information all proceeds to the brain's cortex for analysis. While that's a true statement, it hides a great deal about where it goes before it gets there. Virtually all sensory information first arrives in the brainstem. A slight exception is vision because, if you recall, light activates first the suprachiasmatic nucleus in the hypothalamus (in order to regulate circadian rhythms – see Chapter 3), but after that, fibers from the optic nerve get to the midbrain and brainstem before anywhere else. It's critical to understand that these initial receiving stations deep in the brain aren't passive but are actively engaged in sensation. You'll recall from Chapter 1 that the superior and the inferior colliculus detect light and sound, respectively, and are wired so that they can enable extremely fast behavioral responses, should there be a need to do so. These deep structures have a key role in data analysis. For example, incoming auditory information from the ears is multiplexed (aka muxed), a process in which signals are combined so that they can share the wiring that carries them. The first thing that needs doing by nuclei in the brainstem (notably the cochlear nucleus) is de-multiplexing (aka demuxing). If it isn't done, auditory input will be incomprehensible to the rest of the brain. From these initial sensory stations, information flows through the thalamus, with different parts of the thalamus processing different sensory inputs before sending messages on to the relevant parts of the cortex.

There is a strong possibility here of a common plan: deep brainstem unscrambling and making a first-pass analysis of incoming signals, routing their output through the thalamus, which in turn informs the specialized regions of the cortex. But we used the phrase 'virtually all sensory information' in starting this discussion. Olfaction is different. The olfactory bulbs at the front of the brain receive information from the olfactory mucosa in the nose and send information through the olfactory tracts not to the brainstem but to the amygdala (with some projections to the hypothalamus and thalamus). From here, it goes to the cortex (the piriform cortex and the entorhinal cortex). Perhaps surprisingly, given how closely the two interact, taste information does go to the brainstem, then the thalamus and then the cortex. Why is olfaction organized differently? Bluntly, no one knows, but olfaction is an essential sense; odors can trigger deep memories of all sorts of things. In the novel *À La Recherche Du Temps Perdu* (In Search of Lost Time), Marcel Proust described the vivid reminiscences evoked by the aroma of small cakes – madeleines. Such involuntary memory is now often described as Proustian.

Back in the flow, once sensory information arrives in the cortex, it's subject to yet more detailed analysis, enabling features of the environment (internal as well as external) to be identified. Information is moved around. For example, two cortical visual streams leave the primary visual cortex (aka the occipital cortex). The dorsal stream runs to the parietal cortex and is thought to be involved in guiding actions. The ventral stream goes to the temporal lobe and is more concerned with perception. In simple terms, the dorsal stream deals with 'how' and the ventral stream with 'what'. Vision needs both the guidance of behavior and the identification of objects. This organizational scheme was described in the early 1990s by David Milner and Mel Goodale, and later, it turned out that auditory systems are similarly arranged. From the auditory cortex in the temporal lobe, there are dorsal and ventral streams for dealing with semantic information (the 'what' part) and sound location and movement (the 'where' part).

So sensory information flows through the brain and is assessed and used every step of the way. What about perception? To start with, we could ask, what's the difference between sensation and perception? This question remains a matter of serious scientific and philosophical interest, and many philosophers have written about it. Because we're writing this from an institution whose origins are deep in the Scottish Enlightenment, the work of philosopher Thomas Reid (1710–1796) resonates with us. He was the founder of the Scottish School of Common Sense and recognized that a conscious perception was more than just the sum total of sensation activated. It still works: sensation is the data, perception the extraction of meaning from it, which of course requires the operation of other psychological processes. Perceptions require learning, a bringing-to-bear of past experiences – memories – to understand what current sensory data mean. A remarkable example comes from Project Prakash, founded by Pawan Sinha at MIT. This brilliant scheme corrects visual defects for people in developing countries. Adults, blind from birth because of cataracts in both eyes, have their lenses surgically replaced with new synthetic ones. When light is admitted into the eyes for the first time, peoples' experience is one of confusion, unable to understand what is what or where one thing ends and another starts. Over a few months, people learn to see. Their brain's networks train themselves without supervision so as to make sense of the visual world and to retain that sense as visual memory. They can perceive [13, 14].

So we have input and make sense of it as best we can. But of course, we're bombarded by sensations from within and without. How do we deal with it? It brings us to attention.

ATTENTION

Attention is familiar to everyone. Paying attention in class, paying attention crossing the road, paying attention to the details of a recipe (in the kitchen and in the lab). Naturally, psychologists identify multiple different forms of attention, including the following:

- Selective attention (aka focused attention) describes how we identify and home in on a specific thing to attend to, usually something in the external environment. Other simultaneous events are discounted, though intriguingly, there is evidence that some of the things not apparently attended to can still be recalled later when asked.
- Divided attention: there are limits to how many things can be attended to at once but the ability to process more than one input simultaneously is possible.
- Sustained attention and vigilance: sustained attention describes the continuous performance of a task, such as monitoring a screen looking for discrete events. Sustained attention is often described as vigilance, but it isn't clear that they are wholly synonymous. Sustained attention is a continuous process of watchfulness,

but vigilance can be more discontinuous, a general state of alertness scanning for rare and unpredictable events.
- Joint attention describes the ability to share the point of attention with another person. Pointing to something is a way of doing this but more subtle is following the movement of someone's eyes, sharing their gaze. This is a skill present early in human development so that by the age of around 3 years old, children are fully competent in sharing attention.

A broader way to think is simply in terms of top-down and bottom-up attention. Top-down involves looking for something that has already been identified. In a lab task, one might ask people to select and attend to only blue objects, ignoring anything of another color. Bottom-up attention, on the other hand, is an interruptive process in which focus shifts to something unexpected. These are clearly well established in the animal kingdom. Top-down: think about a crocodile waiting patiently in the water, mostly submerged, paying careful attention to potential food at the water's edge. Bottom-up: think about the accidental crossing of paths between predator and prey.

There is no one single brain mechanism of attention. Different mechanisms exist for top-down and bottom-up because, in the first case, attention is aimed at known and defined events, whereas bottom-up is a rather more 'burglar alarm' like process for early warning. Regions of the cortex, thalamus, basal ganglia, brainstem and cerebellum all have roles in attention. Perhaps most interesting, parts of the cortex and thalamus are thought to have separable roles in the engagement and disengagement of attention, respectively, which is important. When one has fixed attention on something, it has to be possible to actively end the fixation (rather than let attention decay) and move on to something else.

DECISION-MAKING

Decision-making might seem an odd topic to bring into a chapter on cognition, but it isn't. In Chapter 1, we talked about the nerves running through the body. These connect to the various organs of the body and to skeletal muscle, where neuromuscular junctions (which are specialized synapses) act to control muscle fibers. A moment's reflection tells you that so much muscle activity is going to need a prodigious amount of control. The spinal cord is clearly involved, but so too is the brain above. From the cortex through the basal ganglia down to the cerebellum and brainstem – all have a role to play in generating action. However, how are decisions made about what to do and what not to do? This clearly belongs in cognition.

When we talked about the spinal cord (Chapter 1), we noted the existence of reflex arcs capable of causing involuntary yet important movements. There are other kinds of seemingly automatic actions, such as habits: not just nasty habits (like addictions) but any persistent pattern of behavior. Habits are learned, bringing memory into the frame, but once learned, they operate without a lot of conscious thought or effort. Learning to drive a car is a gradual process of transfer from effortful conscious sequencing of actions to doing things habitually. Related, but not at all the same, are what are called motor primitives, small independent patterns of motion from which more complex actions can be built – building blocks if you like. Beyond these are motor programs: established sequences of behavior. Handwriting is a good example because we can separate out two elements. If we gave you pen and paper and asked you to write out the letters of the alphabet, you'd do it. But then if we asked you to grip the pen differently – in a closed fist maybe rather than delicately between your fingers – you could still do it. And in your nonpreferred hand or (with a bit of practice maybe) between your toes. The point is that what you have are two things: a program for the production of letters and an ability to use different muscle groups in service of that program. What you can see here is that movements have a hierarchy, from reflexes through habits to complex programs and deliberate willed actions. But what about decision-making itself, the choosing to do one thing rather than another when there is the choice available? How does that work?

Possibly the most influential theory has been that of Tim Shallice: the supervisory attentional system [15]. It relied on a process he called contention scheduling. Sensory information is processed, and there can be a number of possible actions that follow. The most appropriate response of all those in contention is selected, with other contenders actively inhibited. The basal ganglia are thought to be critical for this process. There is a problem though. Contention scheduling could create an appropriate action, or in the absence of any stimuli, it could either persist with the last action chosen (that is, perseveration), or it could do nothing, putting behavior at a standstill (and if you think this unlikely, spend time watching animals in the wild). Shallice solved this with the idea of a supervisory attentional system, a form of executive mechanism that could bias contention scheduling. It's necessary because the most potent response triggered by sensory input might be inappropriate. You can think of 'inappropriate' in a very moral way – we have ethical principles that bar certain actions – or in a more personal way. Food is a strong stimulus for behavior, but when on a diet, eating is suppressed. The supervisory attentional system was assumed to be able to initiate an action when none would normally be selected. It's an interesting point because there is a sense that while animals, in the absence of relevant stimulation, might do nothing, people will always find things to do.

This psychological model of decision-making has been very influential, with many neuroscientists equating contention scheduling and executive systems with the operations of the basal ganglia (choosing and doing) and prefrontal cortex (overarching executive regulation). There remains of course a question about whether people do actually make the best choices. Unfortunately, the answer is no, not always. Work by Daniel Kahneman and Amos Tversky, crossing economics and psychology, established that people do not always make good choices when their responses are compared to statistical predictions of what would be best. All sorts of reasons sit behind this, to do with (for instance) failing to understand probability, pre-existing beliefs, the numbers of choices available and even the format of material. And of course beyond this, substance abuse can interfere dramatically with decision-making: addiction can almost be described as a disease of the decision-making process.

CHALLENGES

What we've tried to do in this chapter is lay out some ideas that are central to contemporary biopsychological thinking to do with cognition. We chose to focus on memory because it is so fundamental to everything else. Perceiving internal and external events, knowing what to attend to (top-down or bottom-up) and making decisions all rely on the understanding that we have locked in memory. There is more to say about cognition, and some of it will appear in the next chapter, in which we'll talk about the biopsychology of what enables communication between people: emotions and language.

As a last word here, which takes us back to the start, biopsychology, and neuroscience more widely, has operated on the basis that, in the brain, there are separate areas for sensation, cognition and action. It has been a foundational paradigm. Is it a sustainable view? These are our reasons for thinking it is not.

First, we discussed in Chapter 1 the problems of neuroimaging. While fMRI data are consistent within individuals, they are not so across them. Within brain imaging research, there is a crisis of reproducibility (getting consistent results when one data set is reexamined) and replicability (different studies producing comparable outcomes). The problem isn't one of technology: the images are 'real'. The problem is one of expectation, believing that this psychological function ought to be in this place, that one or another.

Second – and deeply relevant to the earlier point – it is becoming increasingly apparent that the way we describe our psychological experiences of the world don't necessarily match physical events in brain systems. Cognition and perception are hard to separate; working memory and attention appear, in brain activity terms, to be similar; emotions and cognition are bound to each other to the extent that it's possible to argue that emotions are in fact forms of cognition. Luis Pessoa describes this as a form of entanglement, cognitive processes deeply entwined and with the brain making what he calls 'collective computations' rather than using a serial system of specialized processes.

Third, adding weight to this but from a different direction: in Chapter 1, we talked about new developments in recording and analyzing the activity of neurons. Several studies show that neurons in the visual cortex respond not only to visual events but to ongoing behavior (and in fact, neurons all over the cortex are driven by current behavior, as well as their more specialist inputs). [16, 17] This blends not just sensation and cognition but cognition and action. What might the point of this be? Probably this: in order to make sense of the world, brains have to understand what changes in the world are caused by the world itself and what changes in its appearance are caused by self-generated movement.

When we move, what we see changes, and if that wasn't factored in, perception would be impossible.

Fourth, there has been a challenge, often unacknowledged in biopsychology, from what is known as 4E cognition. The four Es are embodied (the body affects cognition), enactive (what we do affects cognition), extended (cognition goes beyond the body, such as the use of devices) and embedded cognition (cognition embedded in external events). There is substantial literature around 4E cognition that we don't have the time or space to go into. It is sufficient to say that it can plausibly be argued that cognition involves much more than just events in the brain.

All of this together represents such a challenge to conventional thinking about cognition that we can realistically argue that the old paradigm of sensation–cognition–action working stepwise is breaking down. It makes research and study in this area more exciting than ever.

CHAPTER 5: SUMMARY

- *Cognitive science developed in part as a reaction against behaviorism, enabling a rigorous approach to how we think, understand and interact with the outside world and what's inside us. From their behavior, we infer that other animals show cognition but with differences in, for example, sensory systems based on environmental niche and body form.*
- *Memory, the retention and use of information, is a fundamental cognitive process informing everything we know and do. It is classified into multiple interacting categories. The biological basis of memory is a central research topic; synaptic activity seems to be critical in forming and retaining memories, and in forgetting.*
- *Learning is closely related to memory and, again, is classified into various types, with behaviorist theory informing much of it. Modeling machine learning on brain processes is a significant branch of artificial intelligence.*
- *Sensation across all forms (there are more than five senses) relies on transduction of events into the electrical activity of neurons. Most sensory data arrives in the deep brain and is further processed through the thalamus and cortex. Perception gives raw sensory data meaning based on learning and memory; attention is the targeting of specific sensory events. Perception of events leads to decision-making about what to do.*

- *Challenges in cognition include localization of function — to what extent are cognitive processes separable and confined to dedicated parts of the brain as opposed to networked? What is the relationship between physical events in the brain and psychological experiences? And 4E cognitive theories argue that cognition goes beyond the brain into the body and the world.*

6

EMOTIONS, MOTIVATION AND COMMUNICATION

You can't really divide biopsychology into things that happen internally and things that happen externally because there is an interaction between what happens inside our bodies and how we're seen by others in the world. The previous chapter looked at processes that are primarily internal – how we perceive things, how we pay attention, learn and remember – but now we'll discuss outward-facing events that let other people see what sort of state we're in. Emotions are the main focus, but we'll look at motivation and communication as well. They all tell others about ourselves.

EMOTIONS – HISTORICAL BACKGROUND

> I hate and I love. Why would I do that you might ask. I do not know, but I feel it happening and I am tortured.
>
> Gaius Valerius Catullus (84–54 BCE)

Emotions have fascinated people for a very long time. The depth of history is illustrated by the quote from the poet Catullus and by the writings of Hippocrates, who, as we saw at the beginning of the book, described four humors, blood, black bile, yellow bile and phlegm, each contained in a different part of the body, each associated with a different emotional state. The ancient world recognized the importance of registering emotions through external features too. Aristotle (or more likely someone in his school) wrote a short treatise called *Physiognomonica*, on facial expressions and what they revealed about a person's personality and character. As you might

DOI: 10.4324/9781003215509-7

expect, interest was maintained through the Middle Ages into the Renaissance and Enlightenment and then into the scientific corpus of the nineteenth century, where physiognomy (the systematic examination of faces) sat alongside phrenology (reading bumps on skulls) as a serious way to understand peoples' natures. It's impact was felt in criminology. Sir Francis Galton – whom we also met in the Introduction – created composite photographs that he believed would let him see the facial characteristics of different kinds of people, including criminals. Cesare Lombroso (1835–1909), an eminent Italian physician, went further, using evolutionary theory to argue that criminality was inherent in some people, identifiable by their facial features: large jaws and a low sloping forehead, high cheekbones, hawklike noses and shifty eyes set too closely together. Regrettably, a lot of this sort of nonsense persists, carried over into racial stereotyping. Facial expressions are without a doubt important, but it's all too easy to get carried away.

In 1872, Charles Darwin published 'The Expression of the Emotions in Man and Animals'. He categorized a small number of discrete emotions and focused on facial expressions as registers of emotion, work that has persisted. Darwin was also clear in saying that animals other than humans have emotions, rather inevitably, given his position on evolution. In the twentieth century, the James–Lange and Cannon–Bard theories of emotion reversed what might seem to be the natural order: that emotional experience causes emotional responses. Ideas developed independently by William James and Carl Lange were pulled into a theory that bears their names. James–Lange reversed 'emotional-experience-followed-by-emotional-response' and said instead that bodily changes come first and are interpreted as emotion. But Walter Cannon pointed out that bodily states don't discriminate emotions very well (for instance, changes in heart rate accompany many different emotions); body organs (the viscera) lack sensitivity and are slow to change; artificially induced changes in the viscera do not produce emotion. The body of course contains a brain: the Cannon–Bard theory (Cannon's theories modified by Philip Bard) suggested that sensory systems in the brain, especially the thalamus, were central to understanding emotions.

Stanley Schacter and Jerome Singer looked at these complex interactions in a famous study carried out in 1962. Volunteer subjects were given doses of epinephrine (aka adrenaline) or a placebo. Those given the drug were not told what it was and were then put in different conditions in which they were either unaware of, or misinformed about, the effects of the drug they had been given. The nature of the subjects' responses to the drug depended on emotions being displayed by an experimental stooge – an actor feigning emotion – while the intensity of emotion was highest when subjects were unable to predict what they would experience. The results encouraged a view of emotion that was essentially cognitive – that the physiological change experienced by someone could be interpreted as emotion depending on context (the pretended emotion displayed by the actor) and knowledge (what the subjects thought they knew about the drug). Regrettably, the results were not replicated by other researchers, either by using the same conditions or in different experiments aimed at the same outcome. That is a problem, as the results of the experiment have been persistently quoted despite the lack of further experimental support. There are over 11,000 citations according to the search engine Google Scholar, including very recent ones. Scientific replicability appeared to count for less than people wanting to believe the results.

One more bit of history: as we said in the Introduction, early twentieth-century psychology was dominated by behaviorism. Gilbert Ryle, often described as a behaviorist philosopher, talked about emotions in his famous book 'The Concept of Mind' (1949). He suggested that the word 'emotion' is used to explain behavior in terms of motives or inclinations (as in, "I did it because I was upset"), to describe moods (long-term dispositions – like being a normally genial person) and to describe feelings (transient states that come and go, passing likes or dislikes). Ryle concluded that "emotions are things for which diagnoses are required, not things required for the diagnosis of behavior". This is a challenging statement. He recognized the existence of emotions but dismissed their importance in explaining what people did. All that was needed for that was systematic behavioral analysis (or so it was supposed).

What we see across history are evolutionary, physiological, neurological, behavioral and cognitive approaches to understanding emotions. All of these threads are still woven together through the biopsychology of emotions, as we'll see.

WHAT ARE EMOTIONS?

First, some vocabulary: the word 'emotion' will be familiar; the words 'affect' and 'affective' maybe less so. They're used in psychology, psychiatry and neuroscience quite a lot. Anything 'affective' relates to the emotions, in contrast to cognition or the intellect. 'Affect' is used as a noun, not a verb, to mean a mood or feeling, so someone who is depressed may be described as having 'low affect'. Psychiatrists look at affective disorders; lab scientists study affective neuroscience.

Emotion can be split into three broad dimensions: visceral or physiological (typically indexed by autonomic nervous system activity – heart rate, sweating, pupil dilation and so on); behavioral (emotional display – postures, gestures, facial expressions); and the subjective (felt emotional experiences). The first two are relatively easily measured objectively, the third less so. Felt experiences involve a sense of 'what it is like', which is something at the heart of a great deal of philosophical argument. (The idea of 'what-is-it-like' – the nature of subjective experience – covers a lot of philosophy. Two papers that generated huge interest are [18, 19]. If you look up either of these papers, you will be able to follow long chains of argument and counter-argument through many academic papers about the nature of knowledge and whether everything, including all mental states, is entirely physical.) Despite the simplicity of this triad, there are serious problems in determining what emotions are and how visceral, behavioral and subjective factors interact.

People deliberately or inadvertently transmit signals about their internal states that can be decoded by others. Paul Ekman identified six basic facial expressions of emotion – happiness, sadness, fear, anger, surprise and disgust – that were claimed to be universal human expressions, evolved over time and found across cultures. This idea has been widely promoted in both the scientific and popular press. Paul Ekman has also worked on microexpressions, transient changes in the face lasting only up to four seconds. However, on the basis of sophisticated analytics, Rachel Jack and her colleagues at the University of Glasgow propose just four psychologically irreducible ways to communicate emotion: happy, sad, fear/surprise (approaching danger) and disgust/anger (stationary danger) [20]. Other research has suggested more categories, using combinations of Ekman's original six, and yet others have included love and jealousy with the six, making eight [21]. How many basic expressions there might be is not a settled fact.

The MIRALab at the University of Geneva in Switzerland (http://www.miralab.ch/) ran a project called HUMAINE (Human-Machine Interaction Network on Emotion), which aimed to create computer systems that could record and shape human emotional states and processes. The idea was to make interactions with machines like chatbots more comfortable for humans. However, the point here isn't the artificial intelligence or what the project found. It's the count of 48 emotions in 10 categories, like so:

- **Negative, forceful**: anger, annoyance, contempt, disgust, irritation
- **Negative, not in control**: anxiety, embarrassment, fear, helpless, powerless, worry
- **Negative thoughts**: doubt, envy, frustration, guilt, shame
- **Negative, passive**: boredom, despair, disappointment, hurt, sad
- **Agitation**: stress, shock, tension
- **Positive, lively**: amusement, delight, elation, excitement, happy, joy, pleasure
- **Positive thoughts**: courage, hope, pride, satisfaction, trust
- **Quiet positive**: calm, content, relaxed, relieved, serene
- **Reactive Interest**: polite, surprise
- **Caring**: affection, empathy, friendly, love

There's nothing wrong with this, but other categorizations of emotion include other words, such as admiration, adoration, alarm, avarice, charity, compassion, cruelty, curiosity, dejection, desire, disappointment, discontent, dread, enthusiasm, envy, euphoria, fondness, frustration, greed, grief, hate, humility, hunger, indifference, infatuation, jubilation, miserliness, modesty, panic, patience, pity, rage, regret, restlessness, revulsion, sentimentality, sorrow, sympathy, triumph, woe, wrath and zeal. Some of these fit well into one or other of the ten categories noted earlier, others don't. And some very familiar emotions, such as bittersweet or heartache, are tricky to place.

Does each word describe a discriminable emotion? No, obviously not. They're English language words, which might carry only the subtlest of differences; some are essentially just synonyms. That being the case, can we group them? The point of examining facial expressions was to find 'basic' emotions, but it hasn't produced a definitive conclusion. Other researchers have tried a different approach. In the nineteenth century, Wilhelm Wundt (whom we met in the Introduction) thought that emotions could be assessed on continuums, such as pleasantness and intensity. Like Darwin's belief in evolved basic emotions, Wundt's ideas about dimensions of emotion have also resonated through the centuries.

One prominent current idea is the circumplex model of Lisa Feldman Barrett and James Russell. It has two dimensions: valence (pleasant/unpleasant) and activity (activation/deactivation). Using these two as vectors, it's possible to plot emotions in two-dimensional space. Other theorists, like Robert Plutchik, have developed more complex models also using dimensions. Plutchik believed that there is a small number of basic, primary, or prototype emotions from which others are derived combinatorically. He imagined primary emotions to exist as polar opposites and that they vary in intensity, creating scaled dimensions.

One advantage of using dimensions is that it becomes easier to work across cultures and languages. We're using English language words, but other European languages developed emotion category words English doesn't have. German has *schadenfreude* meaning a malicious enjoyment of the misfortunes of others. While English language speakers may understand the concept, there's not a specific word for it. Another often cited example is in Portuguese: *saudade*, meaning a longing, melancholy or nostalgia. The nearest equivalent in English would be homesickness, something immediately recognizable but which curiously doesn't appear in anyone's list of emotions. Perhaps even more surprising is that Tahitians have no apparent word for 'sadness' or Poles for 'disgust'. (And given that the six basic emotions are supposed to be the products of evolution, this is curious.)

What we're thinking here goes beyond differences in vocabulary across European languages into two different areas. The first is what's called linguistic relativity. This comes under the heading of the Sapir-Whorf hypothesis, named after Edward Sapir (1184–1939) and Benjamin Whorf (1897–1941), who never worked together and never called their ideas 'the Sapir-Whorf hypothesis'. Be that as it may, it's the idea that the structure of language can shape a user's cognition and perception of the world. At its most aggressive, this is a form of linguistic determinism, language strictly shaping thought, but few contemporary linguists take this line. The more common approach is linguistic relativity, language shaping thoughts but not determining or limiting them, though even this is by no means universally accepted. The eminent academic Stephen Pinker rejects the idea outright. Nevertheless, does language help shape emotion? It seems plausible, but a more likely factor affecting emotion is culture. A great deal of psychological research is centered on Western ideas. If we're trying to find out about the human condition worldwide, it's an insufficient perspective for a properly inclusive science. A recent comparative study of emotion looked at a US cohort and another from the Hadza, hunter-gatherers in Tanzania [22]. The study's authors note that

> Hadza descriptions ... [of emotion] ... foregrounded action and bodily sensations, the physical environment, immediate needs, and the experiences of social others. These observations suggest that subjective feelings and internal mental states may not be the organizing principle of emotion the world around.

The words we use and the cultures in which we exist influence and shape so much about us, including our emotional lives. It's at points like this that we can grasp the breadth of intellectual endeavor required to provide complete psychological accounts of complex phenomena such as emotions. And just to press on that even harder, we should think about a couple of philosophical issues that will tie together some of what we've been saying.

Are emotions natural kinds? The idea of a natural kind is a philosophical one. To be very brief, to call something a natural kind is to say that it belongs to a category of things that are all intrinsically the same (whatever surface differences might exist) and different from other kinds of things. The idea of 'natural' comes from their being created in nature: they are not created by people. This philosophical perspective was brought to the front by Lisa Feldman Barrett, distinguished professor at Northeastern University in Boston, Massachusetts. The fundamental question is this: are the universal basic emotions psychologically irreducible? (That is, they can't be broken down any further.) It's the point of Paul Ekman's work on prototypical facial expressions. The alternative is that emotions are not natural kinds but are contextually, culturally and linguistically framed descriptions of our experiences. An increasing volume of work – such as that with the Hadza – points in this direction. It's premature to be definitive, but we believe that the direction of travel favors this non-natural kind of view of emotions. It doesn't discount biology. If we think about disgust as a supposedly prototypical emotion, we immediately see that it has biological utility (don't eat that, it's nasty), but equally, we know that it develops from infancy rather than being innate (very young children will eat their own bodily waste) and 'disgust' has obvious cultural dimensions (some people, on vacation abroad, won't eat local food).

A 'cultural' rather than 'natural kind' approach is concordant with a view of emotions as 'upheavals of thought', drawing them into the sphere of cognition. Martha Nussbaum sees emotions as evaluative appraisals that focus on an object and which require beliefs about that object. If emotions are evaluative appraisals, it not only keeps them in the 'cultural' rather than 'natural kind' camp, it means that they can be thought of as coming under the banner of 'cognition'. Moreover, Martha Nussbaum emphasizes a personal nature to emotions. Because perceptions of internal bodily states are involved in emotions, but no single bodily state defines any emotion, it suggests that any given emotion can be realized in different ways by different people. It moves away from a view of emotions just as uncontrollable urges that have to be tamed.

All in all, what this is pointing at is emotions have biological usefulness, but how we describe our emotions is predicated on context, culture, language and person. However, before you come to your own judgments on what emotions are, we need to look at some more things. How does motivation relate to emotion? And how do brains deal with emotions?

MOTIVATION AND EMOTION

Let's start with three things that might not automatically be thought of as emotions: hunger, thirst and pain. All of these have dedicated sensory systems – look back at Chapter 2. There are physiological and brain mechanisms to detect changes in water balance within the body, and there are mechanisms to detect key nutrients. For some specific hungers, like the need for vitamins, there are no dedicated sensory systems. Selectively deprived of a single vitamin, an animal will learn, if given various diet options, to select the one rich in it. This isn't instant but acquired or learned over a short time based on one particular foodstuff making the animal feel better. Pain has dedicated receptors and sensory pathways. Nociceptors (pain receptors) turn noxious stimuli into electrical activity in A and C fibers that go into the dorsal horn of the spinal cord. From there, pain pathways go into various parts of the brainstem and up into the thalamus and cortex. The proposition is that there are dedicated sensory systems for brains to detect 'important for survival' changes in the state of the body and do something about it, whether physiological adjustment or action in the world.

Hunger, thirst and pain are all forms of primary motivation as distinct from incentive motivation – 'needs' as opposed to 'likes'. For example, eating can come about because of deprivation (needs) or desire, seeing something exceptionally tasty (likes). Motivation isn't the major topic of research it once was. In trying to explain behavior, learning theorists used to talk about drive and central motive states (hypothetical brain machinery that would generate motivation as required), but much of the research on motivation is now concerned with things like decision-making and the control of impulsivity and compulsions (usually in the context of psychopathology). The principal thing we can say is that you can recognize motivated behavior by three characteristics: the choice of action, the strength of that action and the persistence of that action.

Let's pull this back toward emotion. Think back to Chapter 4 on development and the case we cited of the boy Saul. His problem was that there was no systematic relationship between the opportunity to eat and the state of his body's physiology, such that 'hunger' was never labeled properly. And again, earlier on, we spoke about how emotions could be divided into three dimensions: physiological, behavioral and subjective felt experiences. What we have here are basic physiological processes like the need to maintain energy and water levels that we can describe with labels like hunger and thirst, motivational states translating into emotional ones. We could talk about another obvious emotion in the same way – fear, which motivates strong behavioral reactions. We can see fear as a complex state involving underlying physiological changes, cognition (learning and memory in particular) and emotional feelings.

EMOTIONS AND THE BRAIN

The PNS is central to understanding the physical states associated with emotion. Heart rate, pupil dilation, skin conductance, bladder and bowel emptying – they're all part of emotion. Equally, it's accepted that physiological events in the body don't have sufficient specificity to categorize particular human emotions. Can the brain do this? Is there dedicated neural machinery for particular emotions? Maybe specific neurotransmitters, special locations in the brain or particular networks? Back in the day, the existence of such things was widely expected. There's less certainty now. For instance, Martha Nussbaum focuses on grief as a human emotion and is clear in saying that even if we could find loads of people who all processed grief through a particular brain site or system, we would undoubtedly find others who did it differently. Martha Nussbaum is a philosopher, not a neuroscientist. Can we trust her in this? Yes. We'll explain why.

There are three main ways in which the relationships between the brain and emotion have been investigated: experimental studies with laboratory animals, looking at people who have suffered brain damage and through the use of functional neuroimaging, discussed in Chapter 1. Techniques like MRI or CAT scanning, developed in the early 1970s, have enabled detailed examination of the structure of the brain and assisted clinicians with helping people with brain injuries. In a somewhat different space is fMRI: it captures images of blood flow and oxygen use in the brain, which are indicators of neural activity. Some scientists started to raise questions, not about the technology *per se*, but about the interpretation of data. Recently, criticism of fMRI has gone mainstream with doubts about overall reproducibility and replicability – and whether groups of people differentiated by sex, gender, socioeconomic status, education or ethnicity consistently give the same sorts of results (see Chapter 1). The problem is that different people don't necessarily have the same brain areas activated while performing equally well on the psychological tests driving neural activity. Looking for dedicated locations for separate emotions using fMRI is harder than was once expected.

For sure, brains are involved in processing emotion, but we can't be certain that for any particular one it's the same part activated in every person. Is that the end of it? No, of course not. Let's do some more history.

In Chapter 1, we introduced the idea of the limbic system. The word 'limbic' was brought into neuroanatomy by Paul Broca (1824–1880). He described 'la grande lobe limbique', a band of interconnected structures below the neocortex and around the diencephalon. He included the cingulate cortex, hippocampus, amygdala, parts of the hypothalamus (including the mammillary bodies) and some of the thalamus. Other anatomists later added the entire hypothalamus, the septal nuclei, nucleus accumbens and more. Broca confined himself to anatomy, not function. James Papez suggested in 1937 that at least parts of the limbic system were involved in emotion, an idea crystalized by Paul MacLean, who made an explicit association between emotion and the limbic system. Two things remain difficult about the limbic system: first, defining exactly which structures are in it. As well as those listed earlier, others have included the piriform cortex (part of the olfactory cortex), the entorhinal cortex and the fornix (a major fiber bundle connecting the hippocampus with the septal nuclei and mammillary bodies). Whichever particular structures are included, it isn't a self-contained system (that is, only 'limbic' structures connect to each other). All of these diverse parts of the brain connect to other places as well. Following from that is the second question: functionality. Many scientists work to understand what these individual structures do, but it's not easy to identify a common function that unites them all and which would validate the term 'limbic system'. What's more, the word 'limbic' has taken on a life of its own, frequently used by neuroscientists when they just want to highlight something to do with the brain and emotion. Overall, it's not wholly clear that the term 'limbic system' is a particularly useful thing, anatomically or functionally, but here we are, stuck with it. It's persisted partly because, at one time, it all seemed to work.

A number of observations in the 1950s seemed to confirm an association between specific emotions and parts of the limbic system. Some of this makes for grim reading, experiments that would not be performed now. Three things are worth highlighting. In the 1930s, Heinrich Klüver and Paul Bucy found that destruction of the temporal lobes in monkeys produced a raft of behavioral changes that included a reduction in emotional expression. What became known as Klüver-Bucy syndrome was seen in people after surgical destruction of the temporal lobe (which was done for medical reasons in the 1950s). More importantly, Klüver-Bucy syndrome can appear spontaneously, without surgery, as a rare genetic disease. The symptoms can include visual agnosia (an inability to recognize objects), memory loss, dementia and seizures. More pertinently for thinking about the limbic system and emotion, there can be a loss of normal anger and fear responses.

Further experimental work with monkeys in the mid-1950s showed that lesions in the amygdala (which is embedded in the temporal lobe) had effects reminiscent of Klüver-Bucy syndrome, including diminished fear. Also in the 1950s, a phenomenon called septal rage was described, an exceptional level of aggression produced by electrolytic lesions of the septal nuclei. (These lesions were made by passing current down an electrode, insulated except at the tip. The current destroyed all tissue in the vicinity; no one does this anymore.) Lesioned rats would attack and bite objects indiscriminately but would also flee from contact. When they were housed together, they fought incessantly. As we said, grim, but experiments such as these made an association between key parts of the limbic system and changes in behavior, which appeared to confirm that it was central to emotion.

A third discovery in the 1950s – and one that was more intriguing – was that of intracranial self-stimulation. In 1954, James Olds and Peter Milner were interested in turning round the septal rage experiments by stimulating rather than lesioning the septal nuclei [23]. They added a novel twist. Rats had stimulating electrodes implanted under anesthesia into the septal nuclei. Having recovered, they were given control. Working in an operant chamber (a Skinner box – Chapter 5), they could press a lever that would deliver a brief pulse of electricity to the septal nuclei. The rats did it frequently and over long periods of time. What this showed was that while destruction of the septal nuclei produced aggression, stimulating it produced a sensation that rats would actively work to produce. Other species do it too, including humans. It looked like emotion registered in a brain structure, but it actually hasn't turned out to be so simple.

From this dive back into the 1950s, we can point to two things that became most significant for biopsychology: an association between the amygdala, face recognition and emotion, and coming out of intracranial self-stimulation, dopamine and reward. Prosopagnosia is an inability to recognize faces. People with this condition use a variety of other indices to identify individuals. Voice and context, gait, clothing and – rather obviously – hearing someone named are also factors. Prosopagnosia can come about through brain injury or be a developmental (or congenital) condition present from birth. It is not a generalized visual impairment and has multiple forms with variable degrees of face recognition and covert, nonconscious recognition (that is, recognition of a face without awareness). What is still a major issue is the extent to which prosopagnosia is specific for faces or whether it's a more generalized problem of object recognition. Given that we've seen the biological importance some authors give to face processing and emotion, this is a rather polarized argument that, at present, is not fully resolved. However, a recent study of congenital prosopagnosia reviewed all cases of this from 1976 to 2016: 80.3% showed a relationship between impaired face and object recognition, and 19.7% did not [24]. A single study is rarely, if ever conclusive, but this is a striking outcome, with eight out of ten cases showing an association between impaired face recognition and a more generalized object recognition deficit.

The scientific consensus is that face processing uses a part of the temporal lobe called the fusiform gyrus. While this is generally the case, it's not absolute. Interestingly, Marlene Behrmann and Galia Avidan from Carnegie Mellon University, Pittsburgh, examined four individuals with congenital prosopagnosia who showed normal activity in the fusiform gyrus – there was no relationship between their prosopagnosia and this part of the brain. What they did find in these individuals was face-related activation in the frontal lobe which nonprosopagnosic people didn't have [25]. What this highlights is a danger in being absolute when making associations between particular parts of the brain with cognitive and emotional functions. There are inevitably exceptions to the rule.

Nevertheless, the temporal lobe in general and the amygdala (which is embedded in the medial part of it) clearly have a role in face processing. The temporal lobe has high order perceptual functions to help determine what's happening in the world. The amygdala (which is divided into the basolateral and the central nuclei) is likewise associated with complex object recognition – including faces – as well as responding to stress, hedonics (pleasant experiences, like eating) and fear. Indeed, fear is the main emotion with which the amygdala has been associated through many different kinds of studies in humans and in animals. Its connectivity is a clue to its work: as well as receiving highly processed sensory information from the temporal lobe, the amygdala has connections with parts of the basal ganglia concerned with action section; with the hypothalamus, which is actively involved in physiological regulation; and the brainstem, where there are mechanisms for controlling autonomic nervous system activity. This pattern of relationships puts it at an intersection between analysis of what's happening in the world and selecting the best response to it. Those responses involve the choice of action – whether rapid or more considered – and appropriate physiological change. The amygdala's position in this complex circuitry gives it a role in ascribing valence – is something nice or nasty, approach or avoid? – essential for assessing danger and reacting to it.

Fear has been extensively examined in the clinic and the lab, and there is a clear association with the amygdala. Pleasure, on the other hand, has been most closely linked to the neurotransmitter dopamine. This work is predicated on the old studies of intracranial self-stimulation. Two key points became linked. First, intracranial self-stimulation was so rewarding to the animals engaged in it; second, the best sites for the stimulating electrodes were associated with dopamine-contain neurons in the midbrain whose axons projected into the striatum (part of the basal ganglia). Putting these together in the late 1970s led to the claim that dopamine synapses were pleasure synapses. Is that true? No, not totally. There has been a great deal of work looking at the relationship between dopamine activity and states of liking and wanting, as well as anhedonia (the inability to feel pleasure, a significant symptom in depression). Perhaps most important, though, have been experiments looking at what makes midbrain dopamine neurons active. They increase their electrical activity in response to novel positive stimuli, but as the novelty wears off, they become less active. In contrast, their activity goes down if novel aversive stimuli are presented. Again, this activity disappears as the stimuli become more familiar. The most widely accepted interpretation of these effects is this: dopamine neuron activity provides a prediction error signal to the basal ganglia. If something is better than predicted, increased dopamine activity represents a positive prediction error. If something is worse than expected, their activity goes down, representing a negative prediction error. But if something is as expected – whether good or bad – there's no change. Dopamine neuron activity isn't signaling a pure emotion like pleasure but has to do with predictions that enable learning about events in the world.

Overall, what can we say about the brain and human emotions? First and most obvious, you need a brain to have an emotion. The PNS is involved, most definitely, but is not on its own sufficient to account for the behavioral, physiological or experiential elements of emotion. The evidence for specific emotions being tied to specific places in the brain is patchy, and it would be wrong to think in terms of 'fear centers' or 'pleasure centers' or any other sort of center. We can be clear that brains assess the state of the world, that they regulate physiology and that they manage actions, but the systems that do all this are massively networked. We can't simply say that any specific emotion belongs only to one specific center in the brain.

COMMUNICATION

Why do we have this section here? Simple: emotional expressions convey meaning. Putting aside the nature of human-felt experiences and the cultural labeling of emotions, we see what people are doing and what sort of state they're in. On the basis of these observations, we tailor our social interactions. Communication adds to the richness of this social interaction.

Every animal communicates in one way or another: through the secretion of chemicals like pheromones that others detect and interpret and by their actions and vocalizations; birdsong is what most people immediately think of. It's a brilliant example but just one among any number. Other species use frequencies we can't hear. For example, if you record rat colonies, you can bring the ultrasound frequencies they use into our audible range. Rats use an amazing range of calls, conveying meaning just as birdsong does. As you'd expect, nonhuman primates have sophisticated vocal repertoires. Klaus Zuberbühler recorded calls made by Diana monkeys.

Taking the recordings back to his lab, he was able to correlate particular calls with particular actions. Playing the recordings back in natural surroundings confirmed what he thought: calls relating to ground predator threat made monkeys climb up trees, while calls relating to overhead threat made them climb down [26]. There are countless examples of animal communication that you'll have come across or can find out about. But what's of most interest to us here is human communication, which goes way beyond that of any other species. It has complexity and sophistication and powers uniquely rich human cultures.

Before that, we need to step back and look at human nonverbal communication, which is important for our knowledge about and understanding of others. It lacks the formal characteristics of a true language and is neither as stable or reliable as was once thought, and is most certainly not universal [27]. It's often divided into six types, as follows.

1. **Iconic**: these actions work to show what you're talking about, representing, for example, size. A comic trope is an angler using their hands to overestimate the size of their catch.
2. **Metaphoric**: illustrative but abstract, such as one hand going up as the other goes down to indicate that you're weighing something up.
3. **Manipulator**: stroking your chin to highlight thought, opening your mouth and slapping your cheek to show surprise (often ironically), impatient foot tapping, tilting your head to show you're paying attention.
4. **Emblem**: emblems have a meaning of their own. Waving goodbye, thumbs up/thumbs down, gesturing yes and no by head movements.
5. **Deictic**: simple gestures like pointing. Deictic means directly pointing out or demonstrative.
6. **Beat**: these lack meaning but emphasize things like the rhythm and timing of speech. Watch teachers and lecturers carefully, and you'll likely see these.

There is a significant cultural element to gestures. In some countries, nodding your head up and down means yes; in others, it means no; holding the palm of one hand up in some places means 'stop', and in others, it means something vulgar. (The vulgarity itself varies: some countries use one finger for it, others two.) And don't even think about the use of thumb and forefinger to make an 'O'. In some places, it means superb, and in others, it's a crude insult. There are no gestures that can truly be considered universal across human cultures.

What really separates humans from other animals is language. It has features that other forms of communication don't, such as syntax and grammar, a huge, ever-developing vocabulary and a synthetic ability that makes it possible continually to create new written or spoken utterances (and written down or recorded language communicates across time). The classic account of how the brain deals with language highlights Broca's area in the frontal lobe and Wernicke's area in the temporal lobe, connected by a fiber pathway called the arcuate fasciculus. Wernicke's area was supposedly for receiving speech and Broca's area for production, and the left hemisphere was taken to be dominant.

However, while this has been a persistent view of language and the brain, recent work suggests that it is more complex and not as left-lateralized as thought. Studies of speech perception, tracking where sounds originate from, recalling words and their meanings and turning perception into the production of speech have all been shown to use right as well as left hemispheres. Most dramatically, it appears that brain activity relating to words is spread across most of the cortex on both sides of the brain. [28] It's also clear from clinical studies of people who have aphasia (a language communication disorder) that the brain damage that causes it may be located in the cortex, as well as in many subcortical structures, including the basal ganglia, thalamus, cerebellum and even the brainstem. It's clear that auditory information arrives in the brainstem from the ears, and there are remarkable studies showing brainstem responses to speech syllables. The understanding and production of speech are very likely to be more distributed through the whole brain than nineteenth-century neurologists believed.

CONCLUSIONS AND QUESTIONS

Social neuroscience is an area of clear importance in biopsychology, and it cuts across many diverse areas of study. The focus of this chapter has been on two key aspects of social interaction: emotion and communication. How we understand other people (and how they understand us) depends on what is said, what gestures are made, facial expressions and the emotions shown. We're still left with some difficult questions, briefly summarized as follows.

- Some of the conditions that we've described, like hunger, thrust, pain or fear, have obvious biological bases. But does every human emotion we can think of have the same basis in a biological system?
- Although the PNS and the brain have parts to play in understanding and expressing emotions, can we really say that there are dedicated brain regions for emotions?
- Are emotions natural kinds – intrinsically the same across people and with a basis in biology or nature, or do they owe more to learning and culture?

These are big questions that challenge our understanding of what it is to be human. Our hope is that the challenge will not be confrontational but one that stimulates you to research this for yourselves and arrive at conclusions of your own. It might seem paradoxical, but the application of reason will enable us to understand the biopsychology of emotion. Is it worth it? Yes. What we'll deal with in the next chapter are psychopathologies in which disorders of emotion are often present.

CHAPTER 6: SUMMARY

- *Emotions are not straightforward but are identifiable by physiological reactions, changed behavior and felt experience. The circumplex model of emotion uses two opposing dimensions to classify a variety of different ones.*
- *Facial expressions are important for communicating emotion. Paul Ekman identified six universal facial expressions; others disagree on the number, and cross-cultural research suggests that emotions differ and are not natural kinds universally present in all people.*
- *Motivation is related to emotion – motivated behavior is identifiable through the choice of an action, its strength and its persistence. States such as hunger and thirst, fear and reward are understandable in terms of motivation, as well as having emotional content.*

- *The PNS is involved – heart rate, pupil dilation, sweating – but these are nonspecific, cutting across different emotions. In the brain, emotions use the limbic system, amygdala, hypothalamus, temporal lobe, dopamine systems and more – emotions involve many areas rather than being exclusive to one place or another.*
- *Emotions communicate personal states to others. More specifically, nonverbal communication is important and different across cultures. Speech and language are traditionally associated with Broca's and Wernicke's areas, but modern research goes beyond just these. Language, speech and their disorders involve most brain structures in one way or another.*

PSYCHOPATHOLOGIES

Biopsychology tries to shed light on the human condition by looking at how we grow and develop, our cognitive life and emotions. These things are important to know for their own sake, but they become particularly useful in understanding the disruptions in our psychology and behavior known as psychopathologies. We can only scratch the surface in this book but will aim to give you a frame of reference for your own search for more knowledge.

BACKGROUND

The word psychopathology is attributed to Ernst Feuchtersleben, whose work *The Principles of Medical Psychology: Being the Outlines of a Course of Lectures* was translated into English in 1847. Psychopathology covers both psychological and behavioral disturbances, and we use the term in this broad way to cover all of these without regard to whether they are brought about by organic changes within the brain or body or by outside influences. Medical practice deals with psychopathology through psychiatry, a specialism focused on diagnosis, treatment and prevention. Psychology likewise has dedicated areas of specialty: clinical psychology, abnormal psychology, counseling and neuropsychology deal with various problems, from emotional trauma to brain injury. Psychology also takes in learning difficulties, a term that has superseded no longer used pejorative descriptions such as mental subnormality, deficiency, impairment, retardation and handicap.

Psychopathology is universal across human societies. Some cultures have treated people who have psychological problems with compassion, even veneration. Others regard it as a badge of shame to be castigated and hidden. In Western cultures, psychopathologies are treated with a great deal more care than they were maybe 20 years ago or more, but even so, language can be used carelessly. People might talk about something being 'mental' when they only mean 'exciting', or they might say 'mental' and mean it much more negatively – crazy or stupid. It's why we prefer the more formal term 'psychopathology' to mental illness, but even this isn't problem-free. Casual accusations that other people are 'psychopaths' or 'narcissists' are still commonplace.

We'll come to the classification and diagnosis shortly, but in the most general terms, what's involved in psychopathology is shaped by four things. Deviation from accepted standards of behavior (including language and thought), dysfunction in managing one's own life, danger to the safety of oneself and others and distress, both personal and that of others. (Chapter 3 on substance abuse covered some of this.) Perhaps the most problematic is the first one, deviation from what is regarded as acceptable behavior. Times change: in the Western world, homosexuality was a criminal offense until relatively recently (in the United Kingdom, 1967). Worse, homosexuals were subjected to unpleasant forcible treatments because it was regarded as an illness. In the first edition of the American Psychiatric Association's Diagnostic and Statistical Manual of Mental Disorders, published in 1952, homosexuality was included under 'sociopathic personality disturbances'. It was deleted from the second edition in 1973.

A recent concept of relevance to understanding psychopathologies is neurodiversity – related terms are neurotypical and neurodivergent. The concept of neurodiversity originates with Judy Singer, an Australian sociologist. It captures the idea that no two human brains are quite the same. We know that a human brain – or indeed any vertebrate brain – can be divided into a set of structures: hippocampus, amygdala, cerebellum and so on. However, we also saw that the detailed way in which brains wire up, the growth of axons and the development of synapses, is dependent on genomics, physiology and environmental factors. It's inevitable that no two brains will be quite the same in the intricate details of their synaptic connections. Neurodivergence from that which is neurotypical can lead to differences in socialization, communication, perception, learning, emotion and all sorts of other psychological and social functions. Neurodivergence brings difficulties and opportunities that may require specialist help – in education, for example. What's important is that the concept of neurodivergence leads us away from restrictive conceptions of 'abnormality' toward a position in which we see that happy, healthy lives are possible in all sorts of ways for all sorts of people. Neurodiversity is not diagnostic but descriptive. Under its heading, we can find a variety of conditions, which are listed and described in Table 7.1.

Table 7.1 Conditions typically considered under the heading neurodiversity

Condition	Brief description
Autism spectrum disorder	A spectrum because it includes many features, not all of which will be present together. It can include difficulty in communicating and interacting with people and understanding how others think or feel; aversion to bright lights and sounds; anxiety in social situations, especially unfamiliar ones; more time might be required to understand new information; and repetitive and stereotyped behaviors. High-functioning autists perform well intellectually, but at worst, autism is socially isolating and requires considerable support.
Attention-deficit hyperactivity disorder (ADHD)	The principal features are in paying attention, impulsive behavior and overactivity.
Down syndrome	A genetic condition with a full or partial extra copy of chromosome 21 (hence Trisomy 21). It is associated with learning difficulties.
Dyscalculia	More than usual difficulty … with numeracy.
Dyslexia	… with reading.
Dysgraphia	… with writing.
Dyspraxia	… in performing coordinated movements.
Obsessive-compulsive disorder	Obsessive thoughts and compulsive behaviors.
Tourette syndrome	It involves simple tics that are sudden, brief and repetitive, and complex tics involving coordinated movements. Tics can be motor or vocal. Simple vocal tics are, for instance, repetitive coughing or throat clearing. Complex vocal tics can involve repeating phrases over and over or using vulgar and obscene words suddenly and inappropriately.

The idea of neurodiversity has brought out into the open the need for all people to be treated with dignity, tolerance and respect regardless of any label or classification we might assign them. Though Judy Singer coined the term, there is research literature that goes back to the 1950s, when psychologists like Jack Tizard and Anne and Alan Clarke pioneered new approaches to helping people with learning difficulties and the creation of what were then called 'ordinary life models' for individuals who had previously known only institutional care.

CAUSES OF PSYCHOPATHOLOGY

The causes of psychopathology are many and varied. Some conditions are clearly a direct result of altered brain function. Neurological changes caused by neurodegeneration or by trauma, such as a bleed in the brain (stroke) or an external assault on the skull (resulting in a closed head injury), can create the signs and symptoms of psychopathology. Signs are visible evidence of change, and symptoms are a person's descriptions of their experiences. Changes in the brain need not be restricted only to events that start there. Changes in body physiology can lead to disruption of brain function. For example, there is an association between both male and female hormonal activity and schizophrenia. Equally important, some psychopathologies are not related to physical damage but are a reaction to external events in the world, as is the case with post-traumatic stress disorder (PTSD) or reactive depression (aka situational depression).

The question about causality is important because it determines the way in which a problem could be treated. Where there is clear cause and effect, such as a bleed in the brain causing physical damage in a known area, the treatment would be to ameliorate the effects and prevent further occurrences. Unfortunately, the nature of brain changes in disorders like schizophrenia, anxiety and depression are much less obvious. For example, a recent review of 1,340 people at risk for psychosis compared to 1,237 matched healthy individuals (with all participants drawn from many countries) showed no differences in the large-scale anatomy of the brains of healthy people compared to those at risk (including those individuals who went from being at risk to having symptoms) [29]. This detailed analysis confirms what has been seen over and over: no obvious major brain change in psychoses. This doesn't mean that there aren't more subtle physical changes that may cause psychosis. In the past, changes have been reported in the size of the brain's ventricles, white matter volume, altered synaptic pruning and the numbers of neurons and glial cells in specific brain regions (including the cortex, thalamus, basal ganglia, hippocampus, amygdala and cerebellum). Regrettably, no consistent, reliable, or reproducible physical cause has been found. Research into the microstructure of the brain and the operations of neurons is continually expanding, however, and there's still much we do not know about the cause of psychotic illness.

Medical conditions like a bacterial or viral infection lead to a cascade of events that the body will do its best to combat, and if that's insufficient, clinicians can prescribe medications. Likewise, we know that cancerous cells can grow, divide and spread, and that metabolic disorders like diabetes have a clear biological base. There is a logic of cause and effect, a predictable chain of measurable events on which medical practice can intervene. Regular screening can pick up warnings, giving early interventions a much better chance of success, and technologies based on artificial intelligence will undoubtedly improve this even more. Psychopathology isn't like this. Some neurological disorders with overt changes in the brain might have clear indicators, though even here precise definitions can be hard. For instance, Parkinson's disease is closely related to other conditions, such as progressive supranuclear palsy, and there are many other forms of dementing disorder besides Alzheimer's disease. Knowing exactly which one a person has might require an autopsy to define. Worse, psychiatric disorders such as schizophrenia or depression have no clear and obvious biomarkers. There might be underlying biological causes, but at this time, there is no simple test that allows a clinician to define a psychiatric disorder in the way that is possible for very many physical disorders. In 2013, Thomas Insel, then director of the US National Institute for Mental Health (NIMH), captured this dilemma, observing that

> Unlike our definitions of ischemic heart disease, lymphoma, or AIDS ... diagnoses are based on a consensus about clusters of clinical symptoms, not any objective laboratory measure. In the rest of medicine, this would be equivalent to creating diagnostic systems based on the nature of chest pain or the quality of fever. Indeed, symptom-based

> diagnosis, once common in other areas of medicine, has been largely replaced in the past half century as we have understood that symptoms alone rarely indicate the best choice of treatment. Patients with mental disorders deserve better. NIMH has launched the Research Domain Criteria project to transform diagnosis by incorporating genetics, imaging, cognitive science, and other levels of information to lay the foundation for a new classification system. [30]

This was a powerful statement, one that has not gone unchallenged [31]. Thomas Insel's claim follows the line that there are physiological reasons for all psychiatric illnesses, that something has gone wrong with the structure or chemistry of the brain. But are there disorders that are outside this? How should we categorize disorders in which inappropriate learning has taken place, leading to poor psychological and behavioral outcomes?

In practical terms – and we need to be practical because there are so many people who need help – agencies are working together to achieve consistency in diagnosis as a basis for research and understanding. For example, the National Institute of Mental Health in the United States and the Wellcome Trust in the United Kingdom (a major funder of biomedical research) reached an agreement on a common set of measurement packages for mental health and, building on this, the Common Measures in Mental Health Science program [32] is establishing a minimum set of core measures for mental health. The measures in play still have a strong reliance on descriptions of signs and symptoms rather than biomarkers, but this drive toward international consistency is a good step forward.

There is another consideration in regard to psychopathology, one that has attracted a great deal of interest: resilience. Two people exposed to the same adverse life event might react quite differently, with one developing a psychopathological condition and the other not. (This is not to say that people are always either resilient or not, as if it were a personality trait. Individuals might be resilient in one circumstance but not another.) Resilience even extends to neurological conditions. There are now many descriptions of asymptomatic Alzheimer's disease in which people have all the expected pathological markers (physical damage or changes in the brain) but do not exhibit the signs and symptoms of the disorder. Are such people presymptomatic, on the edge of developing the condition? Are there other, as yet unknown pathological markers that they don't have, or perhaps there's some structural brain feature that's protective against the cognitive decline associated with known Alzheimer's pathology. Or are some people's brains more plastic or adaptable in the relationships between brain structure and cognitive and emotional function? As things stand, resilience appears to be present across the different kinds of psychopathology, whatever the cause. It is not clear why, but it does highlight a difference between what we might describe as psychopathologies and other forms of physical illness where cause and effect are more straightforward.

DIAGNOSES OF PSYCHOPATHOLOGY

Let's start with psychosis and neurosis, terms that are familiar and cover a wide range of conditions. The term psychosis refers to conditions in which there are significant changes away from normal thought and behavior but without any significant degree of insight by the people experiencing them. Disorders within the schizophrenia spectrum are still thought of as psychoses, and the medications used to treat them are referred to not only as antischizophrenic but also as antipsychotic drugs. But because the term psychosis covers so many different states serves more as a generic label. The term neurosis derives from the work of Sigmund Freud. It also covers a raft of conditions that affect thought and behavior but where there is no loss of insight, so people suffering from neuroses are aware of the problems they're having. So, psychosis and neurosis are helpful terms but very broad.

There are two widely used diagnostic schemes for psychopathology. One is the ICD-11 – the International Classification of Diseases, eleventh revision [33]. This is owned by the World Health Organization and covers all forms of disease. Its great virtue is in coding, such that the global prevalence of diseases can be monitored. The other widely used scheme is the Diagnostic and Statistical Manual of Mental Disorders (DSM) of the American Psychiatric Association. This is the one on which we will focus here.

It is worth knowing that the signs and symptoms of the various conditions are rarely if ever unique to one particular mental illness or psychopathology. Making a diagnosis – which is of course important in determining what treatment to offer – is a process in which care will be taken to define exactly what signs and symptoms are present, their severity, incidence over time and whether there are hazards to the person or others. It's also important to realize that a lot of the signs and symptoms – possibly all of them – can appear in everyday life. Confusions, delusions, deep sadness, loss of interest, slips of language, feelings of persecution: in some small way, these appear in everyone's daily life. What brings the need for diagnosis and intervention is their frequency and severity, and the difficulties and distress that they cause.

The most recent edition is *DSM-5-TR* (the fifth edition with text revision). Table 7.2 shows the conditions listed, with a brief description of each. It is worth noting that substance abuse may be a trigger for a wide variety of these and that it's important in diagnosis to distinguish between illness that arises from substance abuse from that which does not.

Table 7.2 The diagnostic criteria and codes of *DSM-5-TR*

Diagnostic criteria	Brief description taken from DSM-5-TR
DSM-5-TR has sections on each of these.	The text here is largely as it appears in *DSM-5-TR*, occasionally simplified.
Neurodevelopmental disorders	A group of conditions with onset in the developmental period. They typically manifest early in development, often before the child enters school. Characterized by developmental deficits or differences in brain processes that produce impairments of personal, social, academic, or occupational functioning.
Schizophrenia spectrum and other psychotic disorders	These include schizophrenia, other psychotic disorders, and schizotypal (personality) disorder; abnormalities in one or more of the following five domains: delusions, hallucinations, disorganized thinking (speech), grossly disorganized or abnormal motor behavior (including catatonia) and negative symptoms.
Bipolar and related disorders	These sit between the sections on schizophrenia spectrum and other psychotic disorders and depressive disorders because they bridge between them in symptomatology, family history and genetics. Included are bipolar I disorder, bipolar II disorder, cyclothymic disorder, substance/medication-induced bipolar.

(*Continued*)

Table 7.2 (Continued)

Depressive disorders	Including primarily disruptive mood dysregulation disorder, major depressive disorder, persistent depressive disorder, premenstrual dysphoric disorder. The common feature is a sad, empty or irritable mood, accompanied by changes that affect capacity to function (including somatic and cognitive changes). What differs are issues of duration, timing or presumed etiology.
Anxiety disorders	These share features of excessive fear and anxiety and related behavioral disturbances. Fear is the response to real or perceived future imminent threat, while anxiety is anticipation of threat. Fear is associated with autonomic arousal, thoughts of immediate danger and escape. Anxiety is associated with muscle tension and vigilance in preparation for danger and cautious or avoidance behavior. Panic attacks are a type of fear response not limited to anxiety disorders but present in other disorders as well.
Obsessive-compulsive and related disorders	These include obsessive-compulsive disorder, body dysmorphic disorder, hoarding disorder, trichotillomania (hair-pulling), excoriation (skin-picking) disorder and other specified obsessive-compulsive and related disorders, such as nail biting, lip biting, cheek chewing, obsessional jealousy, olfactory reference disorder.
Trauma- and stressor-related disorders	These are disorders in which exposure to a traumatic or stressful event is listed explicitly as a diagnostic criterion. They include reactive attachment disorder, disinhibited social engagement disorder, PTSD, acute stress disorder, adjustment disorders, and prolonged grief disorder. There is a close relationship between these and anxiety, obsessive-compulsive and dissociative disorders.
Dissociative disorders	These are characterized by a disruption of and/or discontinuity in the normal integration of consciousness, memory, identity, emotion, perception, body representation, motor control, and behavior. They include dissociative identity disorder, dissociative amnesia, depersonalization/derealization disorder, other specified dissociative disorder and unspecified dissociative disorder. Symptoms can disrupt every area of psychological life.

(*Continued*)

Table 7.2 (Continued)

Somatic symptom and related disorders	These have somatic (that is, bodily) symptoms and/or illness anxiety associated with significant distress and impairment. They are more usually seen in primary medical care rather than psychiatric settings.
Feeding and eating disorders	Characterized by a persistent disturbance of eating or eating-related behavior that results in the altered consumption or absorption of food and that significantly impairs physical health or psychosocial functioning. This includes anorexia nervosa, bulimia nervosa, binge-eating disorder and pica (eating things that are not food).
Elimination disorders	These are usually first diagnosed in childhood or adolescence: enuresis is the repeated voiding of urine into inappropriate places; encopresis is the repeated passage of feces into inappropriate places. They may be nocturnal or diurnal (at night or in the day). There are minimum age requirements based on developmental age, not just chronological age.
Sleep-wake disorders	Insomnia, hypersomnolence, narcolepsy, breathing-related sleep disorders, circadian rhythm sleep-wake disorders, non–rapid eye movement sleep arousal disorders, nightmare disorder, rapid eye movement sleep behavior disorder, restless legs syndrome. All involve dissatisfaction with the quality, timing and amount of sleep. Daytime distress and impairment are core features.
Sexual dysfunctions	Sexual dysfunctions are a heterogeneous group of disorders that are typically characterized by a clinically significant disturbance in a person's ability to respond sexually or to experience sexual pleasure.
Gender dysphoria	The area of sex and gender is highly controversial. Here, sex and sexual refer to the biological indicators of male and female, such as in sex chromosomes, gonads, sex hormones, and nonambiguous internal and external genitalia. There are many conditions in which sex development is atypical (such as hermaphroditism). In contrast, gender dysphoria describes the mismatching of sex and gender identity.

(*Continued*)

Table 7.2 (Continued)

Disruptive, impulse control and conduct disorders	Disruptive, impulse control and conduct disorders include conditions involving problems in the self-control of emotions and behaviors. While disruption or impulse control can feature in other conditions, the disorders here are unique because they manifest in behaviors that violate the rights of others (for example, through aggression or destruction of property) and/or that bring significant conflict with societal norms or authority.
Substance-related and addictive disorders	The substance-related disorders encompass alcohol; caffeine; cannabis; hallucinogens; inhalants; opioids; sedatives, hypnotics or anxiolytics; stimulants (amphetamine-type substances, cocaine and other stimulants); tobacco; and other substances. (See the section on addiction in Chapter 3.)
Neurocognitive disorders	These include many neurological conditions such as Alzheimer's disease, vascular dementia, Lewy body dementia and Parkinson's disease, and frontotemporal dementia, traumatic brain injury, HIV infection, Huntington's disease and prion disease. These disorders have a primary clinical deficit in cognitive function that is acquired, not developmental. Although cognitive deficits are present in many, if not all, mental disorders, only disorders whose core features are cognitive are included in this category.
Personality disorders	A personality disorder is an enduring pattern of inner experience and behavior that deviates markedly from the norms and expectations of the individual's culture, is pervasive and inflexible, has an onset in adolescence or early adulthood, is stable over time and leads to distress or impairment.
Paraphilic disorders	These include voyeuristic disorder (spying on others in private activities), exhibitionistic disorder (exposing the genitals), frotteuristic disorder (touching or rubbing against a nonconsenting person), sexual masochism disorder, sexual sadism disorder, pedophiliac disorder, fetishistic disorder and transvestic disorder. They are relatively common but may involve actions that, because of their noxiousness or potential harm to others, are criminal.

(Continued)

Table 7.2 (Continued)

Other mental disorders and additional codes	This section of *DSM-5* provides diagnostic codes for psychiatric presentations that are mental disorders (that is, symptoms cause clinically significant distress or impairment in social, occupational or other important areas of functioning) but that do not meet diagnostic requirements for any of the other mental disorders.
Medication-induced movement disorders and other adverse effects of medication	Medication-induced movement disorders (such as tardive dyskinesia) are included because of their importance in management by medication and because they need to be ruled out in diagnosing other conditions. Although they are labeled "medication induced", it is often difficult to establish the causal relationship between medication exposure and the development of the movement disorder.
Other conditions that may be a focus of clinical attention	This includes conditions and psychosocial or environmental problems that may be a focus of clinical attention or otherwise affect the diagnosis, course, prognosis, or treatment of an individual's mental disorder.

https://doi.org/10.1176/appi.books.9780890425787

Table 7.2 shows us the wide range of psychological and behavioral problems that fall into the remit of psychiatry. In addition, there are neurological problems that also produce psychological and behavioral impairments. Treatment might involve a psychiatrist, but equally neurologists (medics specializing in brain systems), neuropsychologists (psychologists specializing in brain disorders), clinical psychologists (psychologists with a general interest in psychopathology, whether brain related or not) or counseling therapists (offering various types of noninvasive therapy) are likely to be engaged. What sort of conditions are we thinking about here?

- **Traumatic injuries**: these include closed head injuries that happen when force is applied to the head. Being flung to the ground unexpectedly, vehicle crashes, fights – these kinds of things can produce brain injuries immediately or cumulatively over time. Damage can be focal (localized to one place) or diffuse, spread around (as is the case with diffuse axonal injury).

Blood clots can also form. The consequences can be in any psychological or behavioral process and include agnosia (loss of previous knowledge; it can be sensory specific), amnesia (loss of long-term memory or an inability to form new memories), aphasia (loss of language ability) or apraxia (of which four kinds were historically identified: ideomotor apraxia, a problem with simple gestures; ideational apraxia, a problem putting complex actions together; constructional apraxia, a disorder or assembling things; and dressing apraxia, a loss of ability to get dressed). This isn't a complete list, but it covers some of the most significant conditions.

- **Neurodegeneration**: These are slowly progressing disorders in which brain tissue is lost and pathological markers appear (including amyloid plaques, neurofibrillary tangles and Lewy bodies). Neurodegenerative disorders can target brain tissue and white matter – multiple sclerosis, for instance, occurs in various forms, but all involve the stripping away of white matter, compromising axonal function. Psychological and behavioral activities are affected. Alzheimer's disease, for example, principally involves loss of cognitive function, but there are motor effects as well. Likewise, Parkinson's disease and related conditions have clear motor effects but also involve changes in cognition and may be accompanied by depression. Neurodegenerative diseases are associated with various genetic risk factors and only very rarely involve single genes. Huntington's chorea, a progressive disorder of the basal ganglia, is unusual in that it is caused by mutation of only one gene (HTT, which makes a protein called huntingtin that has a role in normal brain function). What triggers abnormal genetic activity and protein formation (or malformation) is not clear: physiological changes, lifestyle factors or exposure to harmful environments can all have a role. The incidence of neurodegenerative disorders is increasing as people live longer.
- **Epilepsy**: About 2% of people will have epileptic episodes. They typically start in childhood and involve tonic/clonic limb contractions (tonic: stiffening of the muscles; clonic: twitching and jerking), loss of consciousness and biting of the tongue, accompanied by involuntary urination. These are called grand mal seizures. In the period after a seizure – the post-ictal period – there will likely be headaches, deep sleep and changes

in mood. Epileptics may also have premonitions of an attack about to happen. Petit mal seizures (aka absence seizures) don't involve overt seizures and are very brief, lasting only a few seconds. Brain electrical activity changes accompanied by temporary loss of consciousness and, frequently, blinking and rolling up of the eyeballs. Many cases are idiopathic (that is, generated spontaneously from within), but there are multiple clinical variants, making precise classification hard. While obviously a problem with brain function, epilepsy is not associated with a particular brain location.

- **Stroke**: This describes the loss of function caused by problems with blood in the brain. Hemorrhagic stroke is a bleed in or around the brain – about 15% of strokes are like this and are the likeliest to be fatal. In contrast to hemorrhage, ischemia is a restriction in blood flow, limiting or eliminating the supply of oxygen to whatever part of the body it occurs in. Ischemic stroke is caused by blood vessel blockage; a transient ischemic attack is the same but brief, with effects often for only a few minutes and never (by definition) more than 24 hours. Aging of brain blood vessels can be a cause, but there are also several associated factors. Hypertension (high blood pressure, known to medics as 'the silent killer') is frequently a cause of stroke, exacerbated by poor lifestyle choices. Indicators of stroke that will make people consult a medic include weakness on one side of the body (hemiplegia), sensory loss on one side of the body (hemianesthesia), visual loss, aphasia, dizziness, double vision (diplopia) or slurred speech (dysarthria).
- **Brain tumors**: these are inappropriate growths in the brain. There will be psychological and behavioral consequences, with the rate of tumor growth related to their severity. Seizures might occur as well. Gliomas are tumors originating in glial cells (astrocytomas coming specifically from astrocytes). Meningiomas are in the brain's meninges, hematomas come from blood vessels and adenomas are in glands – those in the pituitary gland are close to the brain.

Psychopathologies are widespread and varied. They may be congenital, present from before birth, developmental or acquired in later life; they may be caused by organic change or reactions to adverse environments; they can involve medics, clinicians and therapists drawn from many different backgrounds. A huge proportion of people worldwide will, at some point in their lives, need to consult about a psychopathology. What treatments are there?

TREATMENTS FOR PSYCHOPATHOLOGIES

The treatments for psychopathology are many and varied. They can be grouped into physical treatments – pharmaceuticals and surgery – followed by psychological and behavioral approaches.

PHYSICAL TREATMENTS

In Chapter 3, when looking at psychopharmacology, we had an in-depth review of the pharmaceutical treatments used in treating many psychiatric and neurological illnesses. We aren't going to go through these again, but we will note how widely they're used and the considerable success they have had. That isn't to say everything's just fine. As we wrote earlier, a number of treatments were discovered fortuitously rather than designed on the basis of a solid theory about the neurochemistry of disease. L-DOPA treatment for Parkinsonism being an honorable exception. Loss of dopamine was identified as a cause and L-DOPA used to correct it. In contrast the dopamine hypothesis of schizophrenia (that an excess of dopamine is the cause) came about through trying to understand the actions of antipsychotic medications already in use. The same is true for a number of anti-depressive drugs. Nevertheless, while many pharmaceutical treatments have curious histories, they do offer relief if not an actual cure, and treatments in development (such as immunotherapies) have a more robust scientific basis.

There is a strong sense of disapproval when people talk about surgical interventions – psychosurgery. Psychology students the world over will know about the case of the person known in his lifetime only as Patient HM: the late Henry Molaison. In order to treat intractable epilepsy, HM had surgery to remove the hippocampus on both sides of his brain. The operation did alleviate the seizures, but tragically, and totally surprisingly to all concerned, it left him with (among other things) no long-term memory. And then, of course, we have frontal lobotomy, the surgical removal of the prefrontal cortex to treat psychosis. António Egas Moniz, who developed the procedure in the 1930s, won the 1949 Nobel Prize for Physiology or Medicine for this. Nevertheless, there is a sense of moral outrage about frontal lobotomy, which is understandable, if not wholly fair. In the early and mid-twentieth century, awareness of the need for treatments far exceeded any ability to provide them. Frontal lobotomy and other psychosurgeries sit alongside various shock therapies (such as using insulin to induce a coma or drugs to generate convulsions) that one can imagine were well-intentioned but ultimately without merit.

Treatments involving invasive surgery to remove parts of the brain or sever fiber connections are still used in cases where there is resistance to all other forms of treatment. Anterior cingulotomy – making surgically placed lesions in the anterior cingulate cortex – is used to treat chronic refractory depression, pain and obsessive-compulsive disorder. Similarly, anterior capsulotomy is a surgical lesion in the anterior limb of the internal capsule (a fiber system connecting the cerebral cortex and brainstem). Neither of these are widely used; they are not treatments of first choice; and people embarking on them are thoroughly assessed and informed. What remains something of a concern is that neither procedure has a particularly robust theoretical foundation. In contrast, surgical procedures for use in Parkinsonism have a better basis where it is well-known that there is changed activity in many interconnected areas around the basal ganglia. When L-DOPA therapy is no longer effective, surgically placed lesions in part of the globus pallidus (pallidotomy) or thalamus (thalamotomy) can alleviate some of the symptoms of the disorder. These are, however, treatments of last resort.

There are other approaches to the brain rather than making lesions or fiber cuts. Electroconvulsive therapy (ECT) was once widely used, mostly in the treatment of depression and other mood disorders. It is the passage of electric current through the brain, causing seizures. It has been replaced by other treatments and is much less commonly used now, though it might still be made available to treat patients with severe depression or schizophrenic spectrum disorders when all other options are exhausted. ECT is mostly gone, but there are more recent stimulating techniques available.

Transcranial magnetic stimulation (TMS) is what it says: transcranial means across the cranium (that is, the skull), magnetic means that magnetic pulses are generated and stimulation means stimulation. In repetitive TMS (rTMS), an electromagnetic coil is placed against the scalp, and magnetic pulses are sent through the brain. Deep transcranial magnetic stimulation (dTMS) uses a different type of electromagnetic coil that activates deeper parts of the brain. In neither case is there entry of anything other than magnetic waves into brain. Like ECT, it's a noninvasive procedure. Repetitive TMS is used to treat depression that has proved resistant to other forms of therapy and has also been used for obsessive-compulsive disorder. A more recent development, dTMS has been approved by the Food and Drug Administration in the United States for use in treating obsessive-compulsive disorder and smoking. TMS does appear to ease the signs and symptoms of the conditions under treatment. How does it work? Beyond saying that it stimulates the brain, no one can really say. It's safe, but there are often transient side effects: some discomfort to the scalp (almost inevitably), headache, temporary giddiness and unsteadiness, facial tics and spasms, all of which are bearable for people otherwise living with untreatable depression. However, more serious side effects have been seen, including loss of hearing and mania (especially in patients with bipolar disorder rather than depression). It's still too soon to tell if there are long-term effects.

Some procedures are invasive and involve implanting devices in the brain or body. Deep brain stimulation (DBS) was initially developed for use in alleviating Parkinsonism. Electrodes are implanted into the brain and are used to pass current that affects activity in the local region. Control over this is achieved using a pacemaker-like device placed subcutaneously in the chest and wired to the stimulating electrodes; patients have control over it. For Parkinsonism, the favored target was the subthalamic nucleus, a key element of the basal ganglia known to be dysfunctional in the disorder. It is still not wholly clear whether the therapeutic benefit comes from changing the activity of neurons in the subthalamic nucleus or nerve fibers passing close by. Early trials were not universally successful but were sufficiently effective to merit wider use. It has brought benefits to patients for whom standard drug treatments have become ineffective. There was a strong rational basis for the use of DBS in Parkinsonism, but its success has encouraged the use of other basal ganglia disorders such as Huntington's disease, as well as epilepsy, obsessive-compulsive disorder, Tourette's syndrome, chronic pain, cluster headache, dementia, depression, addiction and obesity. Part of what's encouraging such wide-ranging use is the fact that it's low risk, though there are side effects, including seizures, headaches, confusion and lack of concentration, numbness, tingling, balance, speech problems, visual disturbance and mood swings.

A less widely used treatment is vagus nerve stimulation (VNS). An implanted electrode allows patient-controlled electrical stimulation of the vagus nerve. The stimulator is usually implanted in the upper part of the chest, on the left, with a connection to an electrode connected to the vagus nerve in the neck. It's used in epilepsy and depression, typically in patients with severe symptoms resistant to other forms of therapy. How does it work? It appears to trigger activity in the brainstem, which, through the ascending reticular activating system, can change activity over large areas of the brain.

PSYCHOLOGICAL AND BEHAVIORAL TREATMENTS

Behavior therapy (aka behavior modification) is an expression coined by B. F. Skinner and his colleague Ogden Lindsay. Put simply, it's the application of learning theory to the treatment of behavioral disorders. Initially, it was thought possible that psychosis could be treated in this way, but this has not been successful. Neuroses, on the other hand, are responsive to behavior therapy. Depression and obsessive-compulsive disorders are sometimes treated, as are various sexual problems, but perhaps the best known are treatments for anxiety disorders, including phobias. Techniques used include flooding, putting a patient in an inescapable anxiety-provoking situation or (much better) desensitization and being exposed bit-by-bit to anxiety-promoting stimuli or situations. It might start by thinking of (let's say) spiders, then being in the same room as a real caged spider, working toward being able to tolerate being touched by one. Aversive conditioning was briefly used – presenting images of things to be avoided while delivering electric shocks or inducing nausea with drugs – but that has not persisted. Does behavior therapy work? It has been successful with the treatment of simple phobias but not to the extent that behaviorist theoreticians expected. Arnold Lazarus (1932–2013) tried to improve it with multimodal therapy, looking at a more comprehensive package of psychological attributes in addition to behavior: emotion, sensation, imagery, cognition and interpersonal relationships. His work led to a different kind of therapy.

Cognitive behavioral therapy (CBT, also sometimes just cognitive therapy) was developed by Aaron Beck (1921–2021) initially as a treatment for depression, aiming to help people modify their thoughts, which would change the way they lived their lives. Cognition – here meaning thought – was at the heart of it: maladaptive thinking leading to maladaptive behavior. What was not working properly was a cognitive triad: negative views of oneself, one's personal world and the future. It leads to a bleak view of the world characterized by things like selective abstraction (with a negative focus dominating, positive information being discounted), overgeneralization (trivial bits of information being taken out of all proportion) and all-or-none thinking (lacking nuance, giving disproportionate weight to minor events). The role of the therapist is to guide discovery, creating a more logical and rational view of the self and a person's place in the world. It's a therapy that has moved beyond depression into other areas, including anxiety, eating disorders and substance abuse. It can be used on its own or in combination with other therapies. In cases of severe depression, for example, pharmaceutical treatments can help get a person into a state where CBT is more likely to succeed. Does it work every time? No: like any therapy, it works for some people but not others. One of the problems appears to be that, while the therapists who developed CBT relied on the idea that people were in conscious control of their cognition – their thoughts, it turns out that much cognitive processing is automatic and not under conscious control at all, creating potential problems for therapists trying to help clients restructure their cognitive lives. However, all that being said, CBT remains a powerful therapeutic tool.

Rational emotive therapy (aka rational emotive behavior therapy) actually preceded CBT. Albert Ellis (1913–2007) founded it in the 1950s. The theory behind it is that people can be both rational and irrational Being rational is to be constructive, showing both self-help and social utility. Being irrational is the opposite, leading to depression, anxiety and various negative emotions, including shame, blame, self-pity and anger. The role of the therapist is to engage through teaching and examination, encouraging the client to a more rational life. Emotion-focused therapy places emotional life even more squarely at the heart of treatment. It is a more recent development, pioneered by Leslie Greenberg, Robert Elliott, Laura Rice and Sue Johnson. The intention is to help people to make better sense of their emotions in order to overcome their difficulties. Our emotions work as adaptive guides, helping us know what we like, what we should be afraid of and so on. They can become maladaptive through, for example, traumatic experiences in early life. The role of the therapist is to highlight maladaptive emotions and help clients improve their emotional and cognitive lives.

There are many other forms of therapy, including eye movement therapy, dialectical behavior therapy, acceptance and commitment therapy, psychodynamic therapies, psychoanalysis, interpersonal psychotherapy and supportive psychotherapy. We can't go into all of these. The ones we've focused on – behavior therapy, CBT, rational emotive therapy and emotion-focused therapy – all have theoretical bases in psychological theory. Are they properly part of biopsychology? There is no intrinsic regard in any of them for biology. However, they're important. Reliable, fail-safe treatments for psychopathology of whatever sort remain elusive. As such, the best approach is one that keeps all options on the table.

CHAPTER 7: SUMMARY

- *Psychopathology covers many neurological (organic) and psychiatric (mental health) specialisms. The extent to which psychiatric disorders are physical disorders of the brain remains controversial. Neurodiversity is not diagnostic but descriptive, covering a range of conditions that may or may not require specialist help.*
- *Psychopathology has multiple causes, from environmental stressors to internal changes in the brain. The degree to which people might be resilient to stressors or even neurodegeneration is a focus of increasing research.*
- *The diagnosis of neurological conditions is typically based on clear events. Psychiatric diagnosis is based not on physical measures but on signs and symptoms. Many can appear in everyday life: what creates a need for intervention are the frequency and severity of their appearance and the difficulties and distress they cause.*
- *Physical treatments for psychopathologies include pharmaceuticals and various forms of electrical stimulation, such as TMS or DBS.*

Electroconvulsive therapy and psychosurgery are treatments of last resort.

- *Behavior therapy is an effective treatment for conditions such as phobia; cognitive behavior therapy or rational emotive therapy are widely used psychological approaches. There are many other varieties of counseling approaches. What is important is finding the right approach for the right patient.*

8

OVERALL, WHAT DOES BIOPSYCHOLOGY OFFER?

This book is an introduction, not a complete account of biopsychology. There's much more to say and more to know. Some things we've treated in less detail or omitted. But we've tried to cover sufficient material to let you understand something about biopsychology, whether it's a starting point for a career, support for a related subject or just something that piqued your interest.

Now, as a kind of valediction, what can we say about biopsychology? We should begin by reiterating that biopsychology is not a simple exercise in reducing all of psychology to biology. The two interact, and one is not more 'fundamental' than the other. Biological explanations are not of necessity somehow deeper than psychological, or indeed *vice versa*. As an illustration of interaction, we can go back to the Introduction and what we wrote about niche construction, how animals can create and develop their own environments to suit their own needs. It's thought that niche construction can affect evolution. If animals significantly alter their environments, it can change evolutionary selection pressures, making particular traits suited to the niche more useful. All this is very biological, but a line of argument that follows is that human cultures – social organizations, practices, beliefs and so on – are forms of niche construction. It's very plausible but creates a circularity because human cultures feed back into how our cognitive and emotional lives work. Think back to how exposure to what's going on in the world – that is, culture and all that comes with it – generates experience-dependent plasticity, which shapes the development of our nervous systems. Or think again of the discussion we had about how the construction

DOI: 10.4324/9781003215509-9

of emotions is culturally dependent. Biology and psychology need each other: one does not supplant the other.

We also need to reinforce the point that when we talk about human biology, we mean more than just happenings in the brain. Because the brain is so obviously important when we think about biopsychology, we made a lot of effort to look at its structure and the ways in which neurons work. We made an equivalent effort to look at body systems because these are important too. Human beings do so many amazing things that it's easy to forget that the primary drivers of behavior are the need to maintain energy and water levels, to keep warm, to reproduce and to keep the body safe from harm. How brain and body systems communicate, interact and combine to drive behavior is really important. The brain does not sit in splendid isolation from the rest of the body, or indeed from the world around it, both the natural environment and the cultural one.

Thinking about the relationships between brain, behavior and psychology prompted us to ask in Chapter 1, "does the way in which we describe our cognitive and emotional lives, either in terms of everyday language or the more rarefied talk of academic psychology, actually map directly onto the brain? And what does an account of (for example) human memory or perception in terms of neuron firing actually mean? Are we saying that when these neurons in this place are active, we experience something?" These are quite profound questions for biopsychology that we could add an extra twist to.

We often talk about mechanisms — brain mechanisms of perception, attention or memory, for example — but it's not always wholly clear what we mean [34]. At a cellular level, we can see 'mechanism' in a direct causal way. The generation of action potentials is a mechanism behind synaptic transmission, and the binding of a neurotransmitter to a receptor leads through ion channels opening to changes in membrane permeability. 'Mechanism' here is fairly easy to understand: one event causes another through physical means. But when we look at neural systems, ensembles and circuits and try to use their activity as 'mechanisms' to explain how memory or attention are caused, we're on much less solid ground. Activity in neural systems does underlie psychological events, but the relationships between them may be less mechanically causal and more correlative or predictive. Indeed, there is a great deal of interest at the moment in what's called predictive coding, which suggests that the activity of neurons is, in many cases, more about predicting sensory inputs rather than passively reacting to events. Fathoming how psychological states come about through the activity of neurons is a profound problem, made even more complex by theorists who argue for a much more engaged role of the body, external events and culture in shaping psychological life.

And there is of course something that we haven't touched on at all, something that we all know about because we have it: consciousness. It's usual when thinking about consciousness to follow the approach of philosopher David Chalmers and talk about the easy and hard problems. The easy problems aren't just about being asleep or awake but how we deal with events in the world. For example, in dealing with all the stimuli that impact us, how do we understand, separate and integrate across them; how do we attend to them, choosing to focus here and not there; and how do we respond to everything that's happening inside and outside the body? In contrast, the hard problem is to do with our self-aware, subjective conscious experience of 'what-it-is-like'. The nature of consciousness – the hard stuff – tantalizes philosophers and neuroscientists. Is it ineffable, beyond our ability to explain in language or through rational investigation? It might be, but that isn't going to stop anyone attempting to understand scientifically. Is it an emergent property? We can divide emergence into weak and strong categories. Weak emergence is when a complex system has properties that none of its parts have but which can nevertheless be explained – you can't see a picture from a single jigsaw piece, but when they all fit together, there it is. In contrast, strong emergence is the condition when a complex system creates properties that no individual part has but which cannot be explained. Consciousness is often used as an example of this. Nevertheless, neuroscientists are more than willing to try to identify how it is that brain systems give rise to conscious experience.

Several overarching theories of consciousness exist – global neuronal workspace theory, integrated information theory and recurrent processing theory, for example – all of which, in one way or another, believe that neuronal interactions distributed over large areas of the brain create subjective awareness. Tests have been run, but no theory as of yet stands up to detailed interrogation (and moreover, some theories have created extraordinary controversy, with accusations of unverifiable pseudoscience being made). Put bluntly, we're not that close to a detailed and specific account that will resolve the hard problem of consciousness in terms of the combinatorial actions of neurons. Is it worth the effort? Yes, because it's such a key part, not just of biopsychology, but of how we understand ourselves as independent persons. No, because given the manifold number of practical problems that need solutions, spending time and resources on something so out-of-reach can be seen as wasteful. Moreover, there is an argument that care needs to be taken in making claims about the physical nature of consciousness. How we treat and give moral status to other animals – and indeed how we would treat an autonomous artificial intelligence – depends in some large measure on how we assign consciousness to them [35].

Understanding the nature of consciousness is a great challenge. Another, more immediate and pressing, is to understand the nature of mental illness so that we can deliver better prevention and treatment. It is not by any means simple. For example, while different neurological disorders (with clear brain pathologies) have clear genetic risk factors specific to them, psychiatric disorders (where brain pathology is far less clear, to the point of absence) do not. In 2018, the Brainstorm Consortium [36] reported on the results of their examination of multiple studies of more than 200,000 patients with twenty-five different psychiatric disorders. They wrote that "the high degree of genetic correlation among many of the psychiatric disorders adds further evidence that their current clinical boundaries do not reflect distinct underlying pathogenic processes, at least on the genetic level … and that … this suggests a deeply interconnected nature for psychiatric disorders, in contrast to neurological disorders, and underscores the need to refine psychiatric diagnostics". Put simply, what this means is that for neurological disorders where there's an obvious brain pathology, specific genes will be present as risk factors, with different genes for different conditions – Alzheimer's or Parkinson's, for example. But for psychiatric disorders – schizophrenic syndromes, depression, anxiety disorders and so on – there are also genomic risk factors, but it's the same genes across all conditions. At a genome level, there seems to be no discrimination between psychiatric disorders. So how is it that, with the same genomic risk factors, some people have one sort of condition and other people have another?

Understanding the genome will help deal with mental illness but not in any straightforward 'this-gene-that-condition' sort of way. More than biology will be needed, a point reinforced in a recent report by the Royal College of Psychiatrists in the United Kingdom: "With half of mental health conditions established by age 14, there is overwhelming evidence for providing support at the earliest opportunity. ... Ensuring the home environment is free from stress and that children receive the love, attention and care they need is key to protecting their mental health, including in those with a genetic predisposition to illness" [37]. This isn't to say that in childhood, safety first with hugs and kisses solves everything. Rather, it's to say that a stable and predictable nurturing environment is one in which children can be encouraged to learn and grow cognitively, emotionally and socially and develop a sense of self-worth that will enable them to be resilient in the face of life's inevitable chances and changes.

Biopsychology has a lot to offer in a multilayered and complex way: the structure of the brain and its interaction with body systems, how neurotransmission works and how it can be altered by drugs, how the brain develops and the role of experience in shaping it, how our cognitive and emotional lives work and how cultures shape our thoughts and actions. Research that acknowledges the influence of biology, psychology and culture in all of these areas will be needed if the current crisis in mental health is going to be alleviated.

This book highlights the range that biopsychology has. Let's say that you'd like to try, through research, to understand people better and to do something useful. What would you do? Work in neuroscience to understand how the brain forms, how its connectivity is established or how it remodels itself over time? Or maybe work in molecular biology to understand the coding and noncoding genomes better? Or you could train as a research psychologist to improve children's development, maybe focus on their language skills or the development of numeracy and mathematics. Or become a social scientist and try to understand how socioeconomic factors and cultures have biopsychological impact. Or, further along the lifespan, help to understand better the biopsychology of aging and improve the quality of later life for everyone, not just dementia sufferers. There are no limits or restrictions on what you can do as a biopsychologist to help improve, through rational, objective research, the lives of others.

GLOSSARY

Action potential A pulse of electrical activity along an axon (singly or in bursts) initiated in an all-or-none manner at the trigger zone where the axon emerges from a neuronal cell body; also known as a spike.

Activity-dependent neurodevelopment Development caused by neuronal (and glial) activity. An example is the development of a myelin sheath around a neuron; see also experience-dependent neurodevelopment.

Addiction A state in which the presence of a drug or engaging in a particular activity is required for a person to function, even at the expense of long-term harm; see also substance abuse.

Affective disorders Affective relates to affect, meaning emotion. Affective disorders are primarily mood disorders, whether persistent depression or mania.

Amino acids A group of organic molecules: (i) amino acids are the building blocks of proteins and peptides; (ii) human diet: there are eight essential amino acids not synthesized within the body; (iii) amino acids form a class of small molecule neurotransmitters (including glutamate and GABA, gamma-aminobutyric acid).

Amnesia A failure of memory in persons otherwise consciously self-aware: retrograde amnesia – inability to recover past events (after brain trauma, for example); anterograde amnesia – inability to form new memories. Dementing disorders typically feature amnesic syndromes.

Amygdala Brain structure embedded in the lower temporal lobe, internally divided into different areas and with multiple connections in the brain; typically associated with emotion (especially fear), higher-order sensory processing (faces, for example) and memory.

Anxiety Anxiety describes the physiological and psychological responses to threat. It is a normal process but one which can become pathological. Anxiety states include generalized anxiety, a persistent disabling condition; panic attacks, acute states; phobias, irrational fears; and social anxiety, fear of social situations. They can be treated by psychological and/or pharmacological regimes.

Aphasia Impairment in the human capacity for language. There are multiple types of aphasia differentiated by such things as the physical ability to speak, fluency, speech construction, the ability to initiate or repeat speech and comprehension.

Apoptosis and necrosis Apoptosis (aka programmed cell death) is part of neurodevelopment but also present in certain pathological states; it is a natural process in which cells are eliminated and disassembled. Necrosis is not programmed and is a pathological loss of cells.

Artificial intelligence (AI) The intelligence possessed by machines rather than organisms. Computer systems can undertake tasks such as perception, speech recognition and language translation; machine learning uses data and algorithms to improve performance.

Ascending reticular activating system Interacting structures through the brainstem characterized by relatively small numbers of neurons having monoamine neurotransmitters and long axons projecting into every part of the brain and spinal cord. Functionally associated with sleep-wake, alertness, arousal and attention.

Asymptomatic Alzheimer's A condition in which the pathological hallmarks of Alzheimer's disease are present but with minimal psychological impairment. It is a current challenge to explain.

Attention Mechanisms for selecting inputs for examination. Broadly, this can be top-down, deliberately focusing on something chosen, or bottom-up, in which an unexpected stimulus captures attention. There are many specific forms of attention.

Basal ganglia An integrated group of structures, including the caudate nucleus, putamen, globus pallidus, subthalamic nucleus and more. Evolutionarily ancient, concerned with selecting actions when there are options and involved in neurodegenerative disorders, including Parkinsonism and Huntington's chorea.

Behavioral pharmacology The interface between behavioral science and pharmacology, enabling insights into both behavior and the actions of drugs; see also psychopharmacology.

Behaviorism School of psychology focused on the relationships between stimuli (events in the world), responses (actions) and reinforcement of relationships between them. Behaviorists have little interest in mental life (taken to be introspective and not quantifiable) or neural processes; see also black box.

Behavior therapy Behavior therapy developed from behaviorism: a systematic approach to modifying maladaptive or unwanted behavior; successful in treating conditions such as phobia but largely supplanted by other approaches like cognitive therapy.

Behavioral state control We experience three different states: the waking state (moving and consciously self-aware), rapid eye movement sleep (asleep but conscious of dreams) and slow wave sleep (out for the count) with neural and neurochemical differences that need to be regulated; see also sleep.

Black box This refers to a relationship between input and output where no mechanism for transforming one into the other is specified or where it is intrinsically unknowable. Behaviorism views the brain as a black box; artificial intelligence can also have input/output without specified workings.

Blood-brain barrier A physical barrier that regulates transport of molecules from blood vessels into the brain; the ability to cross this is important in drug delivery; see also circumventricular organs.

Brain and ganglia The brain is an interconnected mass of neurons, glia (and other cells) divisible into two hemispheres (right and left), each containing identifiable structures and regions. Ganglia are collections of neurons, glia and other cells but without hemispheres or internal organization.

Brain waves Brains generate electrical activity (measured using scalp electrodes: EEG: electroencephalography). Different waveforms have different frequencies: delta waves (0.5–4Hz), theta (4–7Hz), alpha (8–12Hz), sigma (12–16Hz) and beta (13–30Hz);

they appear at different times and locations and are associated with different functional states. Brain oscillations are rhythmic and/or repetitive spontaneous waves; see also synchronization and desynchronization.

Brainstem That part of the brain above the spinal cord, leading to the midbrain; the cerebellum is above it, intimately wired in. Commonly associated with automatic functions like breathing, it is actually a sophisticated processing area of the brain with sensory, motor and cognitive functions.

Broca's and Wernicke's areas Areas of the left frontal and temporal lobe respectively, identified in the nineteenth century as important for producing and receiving speech (respectively). The control of speech is more complex than this.

Cardiovascular system The heart, blood vessels and blood. There are multiple interactions between the nervous and cardiovascular systems; changes in cardiovascular activity are associated with cognitive, emotional and psychopathological states.

Cell The basic unit of life: eukaryotic cells contain a gel (cytoplasm) in which are embedded structures such as the nucleus, mitochondria, ribosomes and other organelles. Eukaryotic cells are the basis of multicellular life; prokaryotic cells without organelles – unicellular creatures only.

Cerebellum A large structure on top of the brainstem that contains more neurons than any other, including the cortex. Traditionally involved in coordination, it is now recognized that this is not just motor activity but also cognitive – coordination is more than movement.

Cerebrospinal fluid (CSF) Continually produced in the brain's ventricles, CSF bathes neurons, supporting them and helping drain waste away. CSF can be extracted by medics (from the lumbar region of the spinal cord) for analysis in disease states.

Chromosome Threadlike, found in cell nuclei, a chromosome is made of protein and one molecule of DNA. Biological males have X and Y chromosomes; Y is mainly concerned with male sexual features. Biological females have two X chromosomes, one of which is inactivated in every cell during embryonic development.

Circadian rhythms Biological clocks that operate on a 24h cycle and govern physiology and behavior.

Circumventricular organs Those parts of the brain where the blood-brain barrier is absent, enabling direct sampling of the composition of blood.

Coding and noncoding genome A genome is all the genes contained within a body or part of a body (as in a cell genome). The coding genome carries the code for making proteins; the larger noncoding genome has multiple functions to do with gene expression and epigenetics.

Cognitive science A branch of psychology that (unlike behaviorism) scientifically investigates mental states. Heavily influenced by computational theories, it incorporates information theory, linguistics and psychology.

Cognitive therapy Cognitive therapies try to restructure maladaptive cognition through counseling, of which there are multiple forms.

Consciousness Often divided into easy (being awake and alert, processing information) and hard problems (conscious self-awareness). A physical basis for conscious self-awareness has so far proved elusive; whether it is fundamentally unknowable or just waiting to be described is uncertain.

Corpus callosum Major white matter system that connects the right and left hemispheres with each other and with the subcortical brain.

Cortex In Latin, *cortex* means bark, so in biology, cortex refers to an outer part of a structure. The human cerebral cortex surrounds the subcortical tissue below. It has two hemispheres, right and left; in humans, it has substantial sulci (sulcus, grove) and gyri (gyrus, ridge); it has older (allocortex) and newer (neocortex) parts and is typically regarded as critical for higher-order processing.

Cranial nerves The cranial nerves are attached directly to the brain (as opposed to spinal nerves or other nerves of the peripheral nervous system). There are 12, bringing sensory information in and taking motor instructions out.

Critical and sensitive periods Relating to neurodevelopment, a critical period is a window of time in which particular events must happen in order for particular functions to progress; a sensitive period is less intense – a window of opportunity that can recur.

Culture (i) In biology, cultures are collections of cells grown artificially. (ii) Human cultures are systems of beliefs, conventions and principles that bind societies together.

Decision-making For all animals, selecting an appropriate response from competing options is a fundamental computational problem. Impulsion, compulsion or perseveration are potential problems; the basal ganglia are critical for action selection.

Dementia Generic term for an acquired loss of memory, language and other cognitive abilities on a scale that interferes with daily life. Dementia is typically associated with neurogenerative disorders, including Alzheimer's disease.

Dependence In relation to substance abuse, physical dependence is a state in which a drug is necessary for a body to function: unpleasant withdrawal symptoms occur if the drug is unavailable. Psychological dependence is shown through physical and mental health; see also sensitization and tolerance.

Depression A mood disorder that can be severely debilitating and persistent. Duration, timing or presumed cause vary, but a common feature is a sad, empty or irritable mood, bodily and cognitive changes and reduced ability to function as usual.

Diagnostic and Statistical Manual of Mental Health Disorders Widely used diagnostic system specifically for mental illness. Now in its fifth (revised) edition, its introduction in 1952 standardized classification and diagnosis.

Digestive system Those body systems and organs that process food and water. The communication with the nervous system is complex and helps drive behavior appropriate to physiological needs.

DNA (deoxyribonucleic acid) DNA is found in every cell of the body. It has two strands in a double helix connected by nucleotides in base pairs. There are four nucleotides that make base pairs: adenine/thymine, cytosine/guanine; see also coding and noncoding genome, genetics vs. genomics and epigenetics.

Drugs A chemical that has an effect on a biological system. They can be totally synthetic, derived from a natural product or a natural product itself.

Embedded cognition Part of 4E cognition, which sees cognition going beyond the brain: embedded cognition argues that cognition involves continual interaction between a body and the environment (both natural and created).

Embodied cognition Part of 4E cognition, which sees cognition going beyond the brain: embodied cognition emphasizes how cognition involves both bodily state and capabilities – for example, having eight tentacles rather than two arms.

Emergent properties Properties of complex systems that are not properties of individual parts: weak emergence is when properties emerge and how they do so can be explained; strong emergence is when the process is unclear – consciousness is often used as an example.

Emotion Complex states that are typically (though perhaps incorrectly) contrasted to cognitive ones. Emotions are characterized by changes in physiology, in behavior and by felt experience. Many think that emotions are natural kinds; others think that emotions are culturally dependent.

Enactive cognition Part of 4E cognition, which sees cognition going beyond the brain: enactive cognition centers on the idea that cognition comes from the continual interaction of behavior and environment.

Endocrine system The endocrine system synthesizes and releases hormones from multiple organs. Sexual behavior, reproduction, eating, drinking and stress responses are all influenced by the endocrine system; neuroendocrinology is the study of brain-endocrine interactions.

Energy balance Bodies need to take in nutrients, store energy for short- or long-term use, burn energy to maintain temperature and use energy to keep active. Such things are drivers of primary motivated behavior, essential for life.

Enzyme Naturally produced proteins that regulate chemical reactions and metabolic activity; they are involved in building molecules and degrading them. For example, the actions of many neurotransmitters are terminated by specific enzymes.

Epigenetics How DNA is read and how changes in behavior or the environment can influence this. Gene silencing and gene expression – turning genes off or on – are epigenetic mechanisms, changing the activity of genes but not the structure of DNA; see coding and noncoding genome.

Eudaemonia and hedonia Hedonia is the experience of pleasure; anhedonia, the dulling or absence of pleasure, is a key feature of depression. Eudaemonia (aka eudemonia) is a more global state of well-being – being comfortable in one's own skin.

Evolution In biology, the process of change across generations through processes, including natural selection and random mutation; this book does not cover evolution; see Sherrie Lyons, 'Evolution: The Basics'.

Executive mechanism A mechanism that has overarching control of one or more other processes. The prefrontal cortex is often described as having executive functions.

Experience-dependent development Neurodevelopment influenced by what happens to an infant – the social and familial environment, for example, can have significant effects on brain development; see also activity-dependent neurodevelopment.

Explanatory gap A philosophical point: how do chemical and neural events in the brain explain the self-aware experiences a person has? There is a gap between the physical and mental that has not been satisfactorily explained.

Extended cognition Part of 4E cognition, which sees cognition going beyond the brain: extended cognition argues that objects in the world can be part of a cognitive process: smartphones, calculators, abacuses and so on.

Face perception An element of higher-order perception that is important because the face is a key source of information about identity and state; facial expressions have a special place in studying emotion and because of a disorder (prosopagnosia) in which face recognition is impaired.

Fear A behavioral, physiological and psychological state experienced in the face of threat or danger. Normally, it is an appropriate response: various anxiety states result if there is underlying pathology associated with the amygdala and interconnected structures.

Felt experience The state of understanding 'what-it-is-like', which can be emotional, evaluative or intuitive; see also explanatory gap.

Four-E cognition The 4Es are embodied (the body affects cognition), enactive (cognition and action are in separable), extended (cognition goes beyond the body, using devices, for example) and embedded (cognition embedded in external events). They all see cognition as more than a process inside the brain.

Gene expression and silencing Gene expression – a gene is switched on and working: some are expressed all the time, others only at a key time in development, others in particular circumstances. Gene silencing – a gene is switched off; see also epigenetics.

Gene mutations There are many types, some involving risk of disease, some benign. They include single nucleotide variants, single nucleotide polymorphisms and copy number variations. Gene repair is also possible using DNA repair molecules.

Genetics vs. genomics These are not synonyms. Genetics is the study of heredity; genomics is the study of DNA.

Glial cells Critically important for the nervous system: astroglia provide support, oligodendroglia provide myelin (in the brain; Schwann cells do this in the peripheral nerves), microglia work as immune cells and radial glia are critical in neurodevelopment.

Glymphatic system A clearance system in the brain that uses interconnected channels around blood vessels to clear waste material.

Habits Persistent learned patterns of behavior, good and bad: for example, handwashing before eating (good), addiction (not good); can be complex, as in the actions required to drive a car.

Heredity Biological inheritance, passing traits from parents to offspring; sexual reproduction ensures that offspring receive genes from both parents rather than being clones. What is inherited, what is shaped by experience and how they interact remains poorly understood; see also epigenetics, experience-dependent plasticity and genetics vs. genomics.

Hippocampus An important brain structure with functionally discrete areas, neuron types, complex architecture and multiple external connections. It is one of many structures associated with learning and memory (especially spatial and episodic) but has a role in other functions, including emotion and stress.

Homeostasis An important physiological principle – the maintenance (as far as is possible) of a controlled, consistent physiological state; see also energy balance and water balance.

Hormone   Chemical signals synthesized by the endocrine system and released into the bloodstream, traveling long distances to find their targets. They have effects on physical processes (such as sexual reproduction), as well as cognition and emotion.

Hypothalamus   A major structure below the thalamus, toward the bottom of the brain (where it connects to the pituitary gland); divided into discrete nuclei with different connections and functions; it has extensive interconnections throughout the brain and is a key regulator of body physiology.

Immune system   An internal protection system that synthesizes and releases chemicals to deal with inflammation and infection. It uses the lymphatic system, thin tubes throughout the body. Neuroimmunology is an increasingly active area of research.

Incentive – motivation, salience and sensitization   Incentive motivation is motivation driven by learned states (things to like or avoid) rather than by primary needs (like hunger or thirst); incentive salience is the desirability of something; incentive sensitization is an exaggeration of normal incentive salience, important in theories of addiction; see also wanting and liking.

Insomnia   The inability to sleep, a feature in many psycho- and neuropathologies. Onset insomnia, inability to fall asleep; maintenance insomnia, frequent waking; termination insomnia, early waking; drug dependency insomnia, induced by substance abuse.

Ion, ion channel   (i) Atoms – smallest units of matter with all the properties of their element. (ii) Molecule – has multiple bonded atoms. (iii) Ion – an atom or molecule that has a net electrical charge: cations are positively charged, anions negatively. The balance of anions and cations on either side of a neuronal membrane determines electrical charge. Ion channels are pores in a membrane through which specific ions move; regulating their opening and closing is key to neurotransmission.

Knock-out   In biology, knock-out refers to the removal of a particular gene, usually in mice. The opposite is not called a knock-in but transgenic – here a foreign gene is inserted. Knock-out and transgenic mice have enabled significant discoveries in molecular biology and neuroscience. CRISPR (said as crisper, Clustered Regularly Interspaced Short Palindromic Repeats) is a widely used technique.

Language All animals communicate using various means. Human language is qualitatively different, nonstereotyped, with flexible grammar, syntax and vocabulary. How speech is perceived and produced (and aphasia after brain damage) has a long history in neuroscience. It is usually (but not always) left hemisphere dominant. Spoken language is complemented by gestures, a significant part of nonverbal communication; see also Broca's and Wernicke's areas.

Learning A central part of cognitive processing. Unsupervised learning (in people and machines) is that which occurs without training; supervised learning requires structure. Various forms of learning include instrumental (or operant), classical (or Pavlovian), associative and reinforcement learning.

Limbic system First described in the 1800s: a system of connected structures associated with emotion. How viable it will continue to be as an anatomical or functional construct is uncertain.

Mania A state of mental ill-health with overexcitement, delusions, euphoria and hyperactivity. It appears in several conditions; bipolar disorder switches between depression and mania.

Memory Coding, storage, retrieval and use of information — with learning, a fundamental part of cognition. Scientists have an interest in memory consolidation, formation of memory traces (engrams), recall, retrieval and forgetting. A basic division is between working, short- and long-term memory, but psychologists classify many different types; a main goal of neuroscience is to identify precise neural mechanisms of memory.

Meninges Three layers of tissue, separated by cerebrospinal fluid, surrounding the brain: dura mater (which attaches to the skull), the arachnoid membrane and the pia membrane (attached to the brain). The meninges buffer the brain against sudden shocks.

Microbiome All the bacteria in a given location; the gut microbiome, for example, is recognized as important in digestion, including signaling to the brain.

Mirror neurons Neurons activated by a movement or by seeing that movement performed by another; important in neurodevelopment for copying and emulation.

Mitosis and meiosis Mitosis: cell division, each new cell identical to the other. Meiosis: a cell divides twice to produce four cells containing half the original amount of genetic material: it creates gametes (ova and sperm).

Molecular biology The study of molecular organization, including genomics. It has had an impact on neuroscience through the development of tools (such as optogenetics or knock-outs) and insight into the properties of neurons and their interactions.

Monoamine neurotransmitters Monoamines have a single amine group; they include catecholamines (such as dopamine and norepinephrine) and indoleamines (such as serotonin). As neurotransmitters, they tend to be modulatory, are mostly derived from small nuclei in the brainstem, have actions throughout the brain, have many functional properties and are often targets for pharmaceutical therapies in mental illness.

Motivation Sometimes known as drive, the primary motivation is for essentials (food, water and so on), while incentive motivation is driven by learned states (to approach or avoid). Motivation was a major focus for behaviorism; more recently, the focus has been on the role of motivation in addiction.

Motor programs Brains can learn and deploy sequences of coordinated behavior – handwriting, for example. These are more than habits: the program primarily specifies an output – handwriting can be done with either hand, the feet or upside down.

Multiple memory systems It is an accepted principle that different parts of the brain store different kinds of memories; it requires that complex memories are assembled dynamically across these.

Myelin A fatty substance that makes brain white matter white. It is insulation for axons that significantly speeds transmission times; oligodendroglia provide myelin in the brain and spinal cord; Schwann cells do the same for peripheral nerves. Myelination is the deposition of myelin; demyelination is a pathology in disorders, such as multiple sclerosis.

Natural kind A natural kind belongs to a category of things intrinsically the same and different from other kinds of things; they are created in nature, not made by people. A key question relates to emotions: are they natural kinds or culturally determined?

Nerve In biology a nerve in the peripheral nervous system is a collection of neurons nerve cells whose axons are bound together by layers of connective tissue; see also cranial nerves.

Nervous system   It is important to understand that, although it is typically broken down into different parts, there is a unity to the nervous system. It includes the central nervous system (brain and spinal cord) and the peripheral nervous system, which has the autonomic nervous system (divided into sympathetic and parasympathetic), somatic and enteric nervous systems.

Neural network   Neuroscience talks about interconnected neurons as forming a network (or, in smaller instances, an ensemble). Artificial intelligence uses artificial neural networks composed of nodes that simulate the connectivity of organic neurons; such networks can learn, supervised or unsupervised, mimicking natural processes of reinforcement and error correction (backpropagation).

Neurodegeneration   Abnormal and progressive loss of neuronal tissue usually leading to behavioral and mental impairments, typically triggered by genomic events or environmental toxins. Parkinsonism and Alzheimer's are examples; see also asymptomatic Alzheimer's.

Neurodevelopment   The complex progression of molecular and cellular events that create the nervous system. It goes through multiple stages, involves the creation and migration of neurons, formation (and later pruning) of synapses and deposition of myelin. For humans, it ends in late adolescence.

Neurodiversity   Brains develop in relatively predictable ways, leading to neurotypical states; neurodivergence from this can create psychological or behavioral conditions that might require specialist help.

Neuroendocrinology   Study of the relationships between nervous and endocrine systems. Hormone activity has significant effects on the brain in development and adult life, while the brain is a key regulator of body hormone activity; see also endocrine system.

Neurogenesis   The process by which neurons are formed during development. Once thought to be completed during infancy, it now appears that new neurons (adult neurogenesis) can be created later in life.

Neuroimaging   The ability to capture images of the brain. There are multiple techniques for obtaining static images; imaging the activity of the brain using functional magnetic resonance imaging has been especially important for biological psychology.

Neuroimmunology Study of the relationships between nervous and immune systems and how they interact in development, in maintaining normal physiological functions and in response to brain damage; see also immune system.

Neuron (Also spelled neurone, a US/UK difference.) Neurons are the principal cells of the nervous system; glial cells support and enable their functioning; how neurons interact codes information that underpins behavioral and psychological states.

Neuroplasticity The ability of neural tissue to adapt and change in response to events; synaptic plasticity is important in memory formation; see also activity-dependent neurodevelopment and experience-dependent neurodevelopment.

Neurotransmission The process by which neurons release neurotransmitters to communicate with other neurons. Much of this happens at synapses, but nonsynaptic transmission (release of transmitter into the brain where it diffuses to a target) also occurs.

Niche construction How animals can create and develop their own environments; it can affect evolution. It is also argued that that human cultures – social organizations, practices, beliefs – are forms of niche construction.

Nucleus (i) All eukaryotic cells have a nucleus that contains all the chromosomes and genetic material; it is central to cell development and function. (ii) Nucleus also describes a specific part of a larger structure: the hypothalamus and thalamus, for example, are divided into multiple discriminable named nuclei.

Occam's razor When trying to make sense of something, the explanation that requires the smallest number of assumptions is likely to be best.

Osmoregulation Osmosis is the process by which dissolved molecules cross a semi-permeable cell membrane to equalize concentration on either side. Osmoregulation is the process by which water balance in the body's fluid compartments is managed and is important for thirst and drinking.

Phrenology Reading bumps on the skull in order to determine what the brain beneath was like. Nonsense, but it is the origin of the idea of localized psychological functions.

Physicalism A philosophical idea; simply put, it believes that everything is physical, with nothing else required for explaining states or events; see also emergent properties and reductionism.

Predictive coding A theory that brain systems predict sensory events: higher-level sensory areas pass predictions to lower ones, which can generate prediction error signals if there is a mismatch between prediction and actuality.

Prefrontal cortex That part of the frontal lobe at the very front, divided into discrete areas. It is associated with executive control and higher-order cognitive processing.

Primary motivation In contrast to incentive motivation, primary motivation has to do with meeting basic bodily needs: food, water, temperature control and reproduction.

Proteins (i) Composed of 50 or more amino acids, proteins are the basic building blocks of life. Peptides are smaller, made of 2–50 amino acids. (ii) Dietary: proteins are essential nutrients, together with carbohydrates and fats.

Psychedelics Hallucinogenic drugs used recreationally to alter mental states and consciousness. Examples include mescaline, peyote and LSD (lysergic acid diethylamide). Controversially, psychedelics are under review for use in treating mental illness.

Psychopathology Psychological and behavioral disturbance, brought about by unwanted changes within or outside the body.

Psychopharmacology The interaction between psychology and pharmacology: how drugs affect psychological states and how psychological states affect responses to drugs; see also behavioral pharmacology.

Psychosis and neurosis Psychosis – psychopathological conditions in which people experiencing them have no insight; neurosis – conditions where there is insight.

Psychosurgery Surgical intervention in the brain to alleviate psychopathology, used only as a treatment of last resort. Frontal lobotomy is the best known.

Receptors Molecules embedded in neuronal membranes to which specific neurotransmitters temporarily attach like a key in a lock before uncoupling. Neurotransmitter-receptor interaction changes neuronal membrane permeability.

Reductionism A complex philosophical idea; it typically means reducing explanations to simpler levels, but this is not straightforward. For example, biological psychology is not an attempt to reduce all of psychology to biology.

Reinforcement and reward A reward is a positive stimulus or event, but reinforcement is the way in which a reward can shift behavior. Like so: lab rat presses a lever to get a food treat; the food is a reward that reinforces the act of lever pressing.

Replicability and reproducibility Important scientific principles: reproducibility means getting consistent results from repeated independent analysis of one data set. Replicability is coming to the same conclusions using multiple different data sets.

Reproductive cells Reproductive cells (aka sex cells) are more properly called gametes. In biological females, gametes are ova (egg cells); in biological males, gametes are sperm cells. Most human cells are diploid – they have two chromosomes – but gametes are haploid, with one chromosome each. Fusion of gametes leads to new cells with two chromosomes; see also mitosis vs. meiosis.

Resilience The ability to resist, adapt to or quickly recover from negative events, maintaining psychological health. In any given person, it varies rather than being a permanent characteristic. Why some people show resilience is not clear; this is a major focus of research in mental health.

RNA (ribonucleic acid) Unlike DNA, RNA is a single strand and has multiple types. The most important are messenger RNA (a copy of part of a single strand of DNA) and transfer RNA (which drives protein synthesis).

Schedules of reinforcement Important in behaviorism, different types of reinforcement (positive, negative, punishment, omission) and their delivery through various fixed or variable ratio schemes.

Schizophrenia Not a single disorder but a spectrum involving delusions, hallucinations, disorganized thinking (speech), grossly disorganized or abnormal motor behavior (including catatonia) and negative symptoms (which are much like depression).

Sensation and perception Sensation is data entering the nervous system; perception is the extraction of meaning from it, requiring the operation of other processes such as learning and memory.

Senses Exteroceptive senses detect the world outside; interoceptive the world inside the body. Five human senses are normally defined (vision, hearing, taste, touch, smell), but there are more – equilibroception (balance), proprioception, (spatial position) and sensations from body organs. Other animal species have senses for other things – echolocation in bats, for example.

Sensitization and tolerance In relation to substance abuse, tolerance – a reduced reaction to a drug (physical and/or psychological) following repeated administration; sensitization (aka reverse tolerance) – less drug is needed with repeated use to achieve the original effect; see also dependence.

Sleep A complex state essential for normal psychological functioning; dysfunctional in a number of disorders. It divides into rapid eye movement (REM) sleep (aka D sleep) and slow wave sleep (aka non-REM). Electrophysiological recordings highlight differences within and across these; see also behavioral state control.

Spinal cord With the brain, the spinal cord is part of the central nervous system: it has a complex internal organization and runs along the middle of the spinal (vertebral) column (bone); spinal nerves connect to it; it is critical for many sensory and motor functions.

Stimulus-response The heart of behaviorism: the association between stimulus and response, a relationship structured by reinforcement; see also black box.

Substance abuse Because addiction covers a wide variety of forms, substance abuse is the preferred term for addictions relating to drugs and other substances.

Substance dualism Philosophical position (rarely supported) that bodies and minds are separable, with different natures – the body has what the mind does not: it is subject to physical laws and has size, shape, location, solidity and motion; see also physicalism and reductionism.

Synapses Specialized junctions between neurons, with a presynaptic membrane, synaptic cleft and postsynaptic membrane. Synapses are efficient closed systems in which neurotransmission occurs.

Synaptic strengthening Thought to be important for the formation of memories: long-term potentiation is a mechanism that strengthens synaptic activity; long-term depression weakens it.

Synchronization and desynchronization   Brain waves can be synchronized (occurring in a locked pattern) or desynchronized (jumbled). The former is associated with quiet rest and slow wave sleep, the latter with the active waking state and REM sleep.

Thalamus   A large structure in the core of the brain, divided into several discrete nuclei. It has sensory, cognitive and motor functions; connections with multiple parts of the brain; regulates all input to the cortex; and is important in controlling behavioral state.

Transcription and translation   Transcription is making mRNA from DNA; translation is what tRNA does, using the code to make proteins.

Transduction   The process by which sensory events of whatever kind are converted into electrical impulses that travel along nerves.

Vagus nerve   Largest of the cranial nerves (cranial nerve X), with branches throughout the body. Sensory elements carry information from body organs into the brainstem; motor components take instructions back out; it is central to understanding how bodies and brains interact.

Ventricles   Ventricles are chambers: the brain has a ventricular system where cerebrospinal fluid is produced; the heart has ventricles involved in pumping blood.

Wanting and liking   Wanting is the motivation to get something; liking is the enjoyment of it. The incentive-sensitization model of addiction takes wanting to be amplified unnaturally, highlighting addicts' need for drug without necessarily liking it.

Water balance   Bodies need to maintain water balance – taking it in, storing and using it. It is a major driver of primary motivated behavior, essential for life; see also osmoregulation.

White and gray matter   Brains are divided approximately 50:50 into white and gray matter: white is myelinated fibers, gray cell bodies and nonmyelinated axons. There is increasing recognition of the importance of white matter for normal psychological life and involvement in psychopathology.

FURTHER READING

INTRODUCTION

Boyd, R. (editor), *A Different Kind of Animal*. 2018: Princeton University Press.

Strawson, G., *Things That Bother Me*. 2018: New York Review of Books.

Wertheimer, M. and A.E. Puente, *A Brief History of Psychology*. 2020: Routledge.

CHAPTERS 1, 2, & 3

Bear, M.F., B.W. Connors, and M.A. Paradiso, *Neuroscience: Exploring the Brain*. 4th edition. 2016: Wolters-Kluwer.

Carlson, N.R. and M. Burkitt, *Physiology of Behavior*. 13th edition. 2022: Allyn & Bacon.

Poldrack, R.A., *The New Mind Readers*. 2018: Princeton University Press.

Swanson, L.W., *Brain Architecture*. 2003: Oxford University Press.

Uttal, W.R., *Reliability in Cognitive Neuroscience. A Meta-Meta-Analysis*. 2013: MIT Press.

Vanderah, T.W. and D.J. Gould, *Nolte's The Human Brain: An Introduction to Its Functional Anatomy*, 8th edition. 2020: Elsevier.

CHAPTER 4

Davies, J.A., *Life Unfolding: How the Human Body Creates Itself*. 2014: Oxford University Press.

Harris, W.A., *Zero to Birth: How the Human Brain is Built*. 2022: Princeton University Press.

Newton, M., *Savage Girls and Wild Boys*. 2002: Faber & Faber Ltd..

Snowdon, D., *Aging with Grace*. 2001: Fourth Estate.

National Human Genome Research Institute at NIH: https://www.genome.gov/

National Institute of Neurological Disorders and Stroke: https://www.ninds.nih.gov/health-information/patient-caregiver-education/brain-basics-genes-work-brain

CHAPTER 5

Newen, A., L. De Bruin, and S.E. Gallagher, *The Oxford Handbook of 4E Cognition*. 2018: Oxford University Press.

Pessoa, L., *The Entangled Brain*. 2022: MIT Press.

Ward, A., *Sensational: A New Story of Our Senses*. 2023, Profile Books Ltd.

CHAPTER 6

Barrett, L.F., *How Emotions are Made – The Secret Life of the Brain*. 2018: Pan MacMillan.

LeDoux, J., *The Emotional Brain*. 1996: Simon & Schuster NY.

Nussbaum, M.C., *Upheavals of Thought*. 2001: Cambridge University Press. There is a précis available: Nussbaum MC (2004) Précis of Upheavals of Thought. Philosophy and Phenomenological Research 68: 443–449.https://www.jstor.org/stable/40040691

CHAPTER 7

American Psychiatric Association: Diagnostic and Statistical Manual of Mental Disorders, Fifth Edition, Text Revision (DSM-5-TR) https://doi.org/10.1176/appi.books.9780890425787

Insel, T., *Healing: Our Path from Mental Illness to Mental Health*. 2022: Penguin Press NY.

Kring, A.M. and S.L. Johnson, *Abnormal Psychology: The Science and Treatment of Psychological Disorders*. 2022: Wiley.

World Health Organization: International Statistical Classification of Diseases and Related Health Problems. https://www.who.int/standards/classifications/classification-of-diseases

There are any number of specialist support agencies. In the UK the National Health Service (https://www.nhs.uk) and in the USA healthcare providers such as the Mayo Clinic (https://www.mayoclinic.org) have websites that offer comprehensive and reliable guides to psychopathologies and neurological conditions. Do be aware that (i) there are other responsible agencies worldwide; (ii) you should only use websites that are owned and operated by trustworthy agencies: there is a lot of low quality information in the web; and (iii) the web is no substitute for actual consultation with a clinician.

REFERENCES

1. Winston, P.H., The next 50 years: A personal view. *Biologically Inspired Cognitive Architectures*, 2012. **1**: pp. 92–99.
2. Hobson, H. Animal research statistics for Great Britain. Available from: https://www.understandinganimalresearch.org.uk/news/animal-research-statistics-for-great-britain-2021
3. Segal, D.R., *Drug Use in the U.S. Army*, University of Maryland, Editor. 1976, Jimmy Carter Presidential Library and Museum.
4. Huguet, G., et al., Genome-wide analysis of gene dosage in 24,092 individuals estimates that 10,000 genes modulate cognitive ability. *Molecular Psychiatry*, 2021. **26**(6): pp. 2663–2676.
5. Bruch, H., *Eating Disorders: Obesity, Anorexia Nervosa, and the Person Within*. 1979. Basic Books.
6. Esteban-Cornejo, I., et al., Paediatric obesity and brain functioning: The role of physical activity-A novel and important expert opinion of the European Childhood Obesity Group. *Pediatric Obesity*, 2020. **15**(9): p. 4.
7. Bethlehem, R.A.I., et al., Brain charts for the human lifespan (vol 604, pg 525, 2022). *Nature*, 2022. **610**(7931): pp. E6–E6.
8. Willroth, E.C., et al., Well-Being and cognitive resilience to dementia-related neuropathology. *Psychological Science*, 2023. **34**(3): pp. 283–297.
9. Siegel, J.M., et al., A brief history of hypocretin/orexin and narcolepsy. *Neuropsychopharmacology*, 2001. **25**: p. S14–S20.
10. Cooke, J.R. and S. Ancoli-Israel, Chapter 41 - Normal and abnormal sleep in the elderly, in *Handbook of Clinical Neurology*, P. Montagna and S. Chokroverty, Editors. 2011, Elsevier. p. 653–665.
11. Bartol, T.M., et al., Nanoconnectomic upper bound on the variability of synaptic plasticity. *Elife*, 2015. **4**: p. 18.
12. Ffytche, M., *Sigmund Freud*. 2022: World Rights: Reaktion.
13. *Project Prakash*. 2017 12-03-24]; Available from: https://www.projectprakash.org/

14. Chatterjee, R. *Giving blind people sight illuminates the brain's secrets*. 2015 12-03-24]; Available from: https://www.science.org/content/article/feature-giving-blind-people-sight-illuminates-brain-s-secrets
15. Shallice, T., *From Neuropsychology to Mental Structure*. 1988: Cambridge University Press.
16. Stringer, C. et al. Spontaneous behaviors drive multidimensional, brain-wide activity. *Science*, 2019. **364**(6437): pp. 1–11.
17. Musall, S. et al. Single-trial neural dynamics are dominated by richly varied movements. *Nature Neuroscience*, 2019. **22**: pp. 1678–1686.
18. Nagel, T., What is it like to be a bat. *Philosophical Review*, 1974. **83**(4): pp. 435–450.
19. Jackson, F., What Mary didn't know. *The Journal of Philosophy*, 1986. **83**(5): pp. 291–295.
20. Jack, R.E., O.G.B. Garrod, and P.G. Schyns, Dynamic facial expressions of emotion transmit an evolving hierarchy of signals over time. *Current Biology*, 2014. **24**(2): p. 187–192.
21. Sabini, J. and M. Silver, *Ekman's basic emotions: Why not love and jealousy? Cognition & Emotion*, 2005. **19**(5): pp. 693–712.
22. Hoemann, K., et al., What we can learn about emotion by talking with the Hadza. *Perspectives on Psychological Science*, 2024. **19**(1): pp. 173–200.
23. Olds, J. and P. Milner, Positive reinforcement produced by electrical stimulation of septal area and other regions of rat brain. *Journal of Comparative and Physiological Psychology*, 1954. **47**(6): pp. 419–427.
24. Geskin, J. and M. Behrmann, Congenital prosopagnosia without object agnosia? A Literature Review. *Cogn Neuropsychol*, 2018. **35**(1–2): pp. 4–54.
25. Behrmann, M. and G. Avidan, Congenital prosopagnosia: Face-blind from birth. *Trends in Cognitive Sciences*, 2005. **9**(4): pp. 180–187.
26. Zuberbühler, K., Predator-specific alarm calls in Campbell's monkeys, Cercopithecus campbelli. *Behavioral Ecology and Sociobiology*, 2001. **50**(5): pp. 414–422.
27. Patterson, M.L., A.J. Fridlund, and C. Crivelli, Four misconceptions about nonverbal communication. *Perspectives on Psychological Science*, 2023. **18**(6): pp. 1388–1411.
28. Huth, A.G. et al. Natural speech reveals the semantic maps that tile human cerebral cortex. *Nature*, 2016. **532**: pp. 453–458.
29. Haas, S.S., et al. Normative modeling of brain morphometry in clinical high risk for psychosis. *JAMA Psychiatry*, 2024. **81**(1): pp. 77–88.
30. Insel, T. Transforming diagnosis. 2013; Available from: https://psychrights.org/2013/130429NIMHTransformingDiagnosis.htm
31. Hickey, P. Transforming diagnosis: The Thomas Insel article. 2013; Available from: https://www.behaviorismandmentalhealth.com/2013/05/05/transforming-diagnosis-the-thomas-insel-article/

32. International Alliance of Mental Health Research Funders, *Driving the adoption of common measures*. Available from: https://iamhrf.org/projects/driving-adoption-common-measures
33. World Health Organization, *International Classification of Diseases 11th Revision*. Available from: https://icd.who.int/en
34. Ross, L.N. and D.S. Bassett, Causation in neuroscience: Keeping mechanism meaningful. *Nature Reviews Neuroscience*, 2024. **25**(2): pp. 81–90.
35. Mazor, M., et al., The scientific study of consciousness cannot and should not be morally neutral. *Perspectives on Psychological Science*, 2023. **18**(3): pp. 535–543.
36. Anttila, V., et al., Analysis of shared heritability in common disorders of the brain. *Science*, 2018. **360**(6395).
37. Royal College of Psychiatrists, *Infant and early childhood mental health: The case for action*. 2023; Available from: https://www.rcpsych.ac.uk/improving-care/campaigning-for-better-mental-health-policy/college-reports/2023-college-reports/infant-and-early-childhood-mental-health--the-case-for-action-(cr238)#:~:text=In%20this%20landmark%20College%20report,get%20the%20support%20they%20need

INDEX

Pages in *italics* refer to figures and pages in **bold** refer to tables.

action potential 61–63, 67, 75, 97, 163, 168
addiction 60, 78–79, 117–118, 158, 169, 176; drug dependence 71–72, 173; drug tolerance and sensitization 71–74, 184; incentive-motivation 177, 179
adipose tissue 53–54, 57, 90
adrenaline 60, 64–65, **77**, 124
aging 80, 83, 92–95, 98, 100, 111, 154, 166
allocortex 36, 44, 172
anhedonia 135, 175
anorexia nervosa **150**
Alzheimer's disease **77**, 111, 145, **151**, 153, 165, 173, 180; asymptomatic 94–95, 146–147, 169; genes 94; immunotherapy 75–76; inflammation 50
amygdala 35–36, 140, 142, 145, 168; emotion 107, 132–135, 175; olfaction 114
anxiety 30, 71, 126, 144, **149–150**, 165, 169; neurodiversity **143**; treatment **77**, 158–159
apoptosis 93, 169
aphasia 138, 153–154, 169
apraxia 153, 178
Aquinas, Saint Thomas 3
Aristotle 3, 8, 10, 91, 112, 122
artificial intelligence 8, 88, 103, 120, 126, 145, 165, 169–170, 180
attention 57, 99, 116–120, 137, **143**, 152, 163, 169; divided 115; joint 116; selective 115; sustained 115

autonomic nervous system 27–28, 44, 125, 135, **149**, 180
Avidan, Galia 134
axon 17, *18*, 20–22, 61–63, 67, 78, 86–88, 100, 152, 168; diffuse axonal injury 152

bacteria 9, 23, 50, 52, 145, 178
Bard, Philip 123
Barrett, Lisa Feldman 127–128
basal forebrain 35, 44
basal ganglia 31, *32*, 44, 145, 169; decision making 34–35, 73, 116–118, 135–136, 173; Huntington's chorea 153; Parkinson's disease 75, 156–157; speech 138
Beck, Aaron 159
behaviorism/behaviorist 8, 102, 108–110, 120, 124, 158, 170, 172, 179, 183–184
Behrmann, Marlene 134
Bernard, Claude 7
Berridge, Kent 73
Bessel, Friedrich 5, 9
bile 4, 122
bipolar disorder **77**, **148**, 157, 178
Bliss, Tim 106
blood-brain barrier 23, 48, 56, 68, 170, 172
brainstem 30–35, 44, 46, 56, 86, 94, 99, 156, 171; action 116–117; ascending reticular activating system 76, 97, 169, 179; peripheral nervous system 27–28, 53, 135, 185;

sensation and attention 113–114, 116, 130; speech 138; vagal nerve stimulation 158
brain surgery 36, 133; lobotomy 14, 155–156, 182; pallidotomy 156; psychosurgery 14, 36, 155, 161, 182; thalamotomy 156
brain tumors: adenomas 154; astrocytomas 154; gliomas 154; hematomas 154; meningiomas 154
Broca, Paul 35–36, 132, 138, 140, 171, 178
Brodmann, Korbinian 38
brown adipose tissue (brown fat) 54, 90
Bruch, Hilde 89–90
Bucy, Paul 132–133

Cajal, Santiago Ramon y 7
Cannon, Walter 7, 123
cardiovascular system 9, 16, 47, 48, 58, 60, 171
Catullus, Gaius 122
central nervous system (CNS) 16–17, 22–27, 44, 58, 180, 184
cerebellum 31–33, 37, 39, 44, 86, 142, 145, 171; attention 116; decision making 117; procedural memory 107; speech 138
cerebrospinal fluid (CSF) 16, 22, 54, 171, 178, 185
Chalmers, David 164
chromosomes 82, 94, 100, **143**, **150**, 171, 181, 183
circadian rhythms 70, 99, 113, **150**, 171
Clarke, Alan 144
Clarke, Anne 144
cognition 10, 44, 59, 112, **151**, 153, 173, 177–178, 182; challenges 118–121; definition 101–103; embedded 120, 173; embodied 120, 174; emotion 125, 128–129, 131; enactive 58, 120, 174; extended 120, 175; four Es 120, 175–176; genetics 87; sleep 95; therapy 159, 172

communication 8, 48, 63, 119, 122, 136–140, 142, 173, 178
computerized axial tomography (CAT) 41–42, 131
consciousness 164–165, 172, 174, 178
Copernicus, Nikolaus 101
corpus callosum 30–31, 32, 39, 172
corticostriatal loops 35, 43
critical periods 21, 88, 172

Darwin, Charles 7, 12–13, 68, 101, 123, 127
decision making 8, 59, 69, 73, 116–118, 120, 130, 173
deep brain stimulation 42, 157
Dement, William 98
dementia 50, 58, **77**, 133, **151**, 158, 167, 173
dendrites 17, 18, 61–63
depression 31, 58, 71, 153, 165, 168, 173, 175, 178, 183; anhedonia 135, 175; causes 144–145; treatment 42, 58, **77–78**, 156–159
Dews, Peter 69
diffusion tensor imaging 42
digestive system 50–52, 58, 173
DNA 2, 52, 80–85, 87, 100, 102, 106, 171, 173–174, 176, 183, 185
Donne, John 107
dorsal stream 114
drugs 60, 66–70, 78–79, 83, 98, 124, 156, 166, 173, 177, 182; addiction 70–74, 168, 185; therapeutic 74–76, **77**, 78, 147, 155, 158, 170, 173, 182, 184
dualism 11, 108, 184

Ekman, Paul 125–126, 128, 139
electroconvulsive therapy (ECT) 156–157, 161
electroencephalography (EEG) 41, **96**, 170
Elliott, Robert 160
Ellis, Albert 159
emergent properties 12, 87, 164, 174

emotions 4, 8, 30, 58, 90, 119, 139–141, **151**, 159–160, 168, 174, 179; and the brain 35–37, 131–136, 178; categories 125–129; definitions 125–126; historical background 122–124; motivation 129–131
endocrine system 48, 54, 58, 91, 174, 177, 180; *see also* hormones
energy balance 53, 174, 176
Enlightenment, The 4, 9, 114, 123
enteric nervous system 28, 44, 180
epigenetics 83–85, 87, 100, 102, 106, 172, 174, 176
epilepsy 36, 153–155, 157–158
epinephrine *see* adrenaline
eudaemonia 95, 175
eugenics 13–14
evolution 7, 12, 34, 80–81, 84, 101, 123–124, 127, 162, 175, 181
executive system 118, 175, 182
explanatory gap 44–45, 108, 175

fear 35, 125–126, 130, 133, 135–136, 139, **149**, 169, 175
Feuchtersleben, Ernst 141
forgetting 111, 120, 178
Freud, Sigmund 111, 147
frontal lobe 35, *37*, 38, *39*, 93, 134, 138, 182

Galen 4–5, 15
Gall, Francis 6, 9, 43
Galton, Francis 13, 123
ganglia (ganglion) 27–28, 46, 170
genes 20, 31, 56, 81–85, 94–95, 100, 106, 165–166, 177; brain development 21, 84–85, 90; circadian rhythms 70; diseases 94, 153, 165, 172, 174, 176
genetics 2, 13–14, 17, 21, 40, 80–82, 87, 100, 102, 133, **143**, **148**, 153, 165–166, 176, 179, 181
genomics 31, 43, 80–83, 100, 102, 142, 165–166, 173, 176, 179

glia/glial 44, 50, 85, 88, 100, 145, 154, 168, 170, 181; oligodendroglia 20–21, 62, 86, 176, 179; astroglia 20, 23, 106, 176; microglia 20, 176; radial 20, 85, 176
glymphatic *49*, 50, 176
Goodale, Mel 114
gray matter 5, *24*, 25, 30–31, *32*, 33, 44, 92, 185
Greenberg, Leslie 160

Harvey, William 5
heart *see* cardiovascular system
Hebb, Donald 106
hedonia *see* eudaemonia
hemisphere(s) 27, 30, 44, 70, 99, 138, 170, 172, 178
heredity 102, 176
hippocampus 35–36, 42, 104–107, 142, 145, 155, 176; neurogenesis 86, 88; aging 93; episodic memory 104; spatial memory 105; long term potentiation/depression 106–107; limbic system 132
Hippocrates 4, 9, 12, 15, 30, 57, 122
homeostasis 7, 176
hormones 48, 54–57, 63, 81, 91, 99, **150**, 174, 177, 180
humors 4, 12, 30, 57, 122
hunger 48, 126, 129–130, 139, 177
Huntington's disease 94, **151**, 153, 157, 170
hypothalamus 31, 34–36, *39*, 44, 53–56, 70, 97, 113–114, 132, 135 140, 177

immune system 9, 48, *49*, 58, 177, 181; histamine 64; immunotherapy 75–76; microglia 20; neuroimmunology 50; neuroinflammation 50; radiotherapy 93
inflammation 49–50
Insel, Thomas 145–146
Itard, Jean 6, 9

Jack, Rachel 125
Johnson, Sue 160
junk DNA 83
Juvenal 93

Kahneman, Daniel 118
Kepler, Johannes 101
Kinnebrook, David 5
Klüver, Heinrich 132–133
Kuhn, Thomas 101

Lange, Carl 123
language 43, 85, 102, 142, 148, 164, 166, 178; acquisition 88, 90; brain 138, 140; emotion 119, 126–129, 163
Lazarus, Arnold 158
learning 15, 35, 58–59, 105, 115, 117, 120, 131, 146, 176, 178, 183; addiction 72–73; artificial intelligence 169; behaviorism 7–8, 69; behavior therapy 158; development 87–90; dopamine 136; emotions 139; learning and memory 108–111; neurodivergence 141–142, **143**, 144; schedules of reinforcement 8, 33, 109–110, 170, 178, 183–184
Levitis, Daniel 2
limbic system 31, 36–37, 40, 132–133, 140, 178
liver 51–53, 67
Lindsay, Ogden 158
Lombroso, Cesare 123
Lømo, Terje 106
Loyola, Saint Ignatius 91
Lyon, Mary 82
Lyon, Mel 69
Lysenko, Trofim 84

MacLean, Paul 132
magnetic resonance imaging (MRI/fMRI) 42–43, 119, 131, 181
mania **77**, **149**, 157, 168, 179
Maskelyne, Nevil 5
medulla oblongata 31, *32*, *39*

memory 69, 102–103, 112, 114–115, 120, 131, 133, **149**, 169, 173, 178, 181, 183; amnesia 36, 111, **149**, 153, 155, 168; autobiographical 104; biological basis 35–36, 58–59, 87, 93–94, 105–107, 176, 179; classification 104–105; declarative 104–105; episodic 104, 176; habits 117–118; neuronal firing 44, 66, 163; learning 108–110; procedural 105, 107; semantic 104, 114; sensory 105; sleep 99; spatial 36, 105, 107, 176; working 103, 105, 107, 119
meninges 22–23, 36, 48, 154, 178
microbiome 52, 178
midbrain 31–34, 44, 97, 135, 171
Milner, David 114
Milner, Peter 133
mirror neurons 89, 178
Molaison, Henry (HM) 36, 155
Moniz, Egas 14, 156
Morris, Richard 106
Moser, Edvard 107
Moser, Mary-Britt 107
motor cortex 38
multiple sclerosis 50, 86, 153, 179
myelin and myelination *18*, 21–22, 25, 30, 62, 86, 88, 93–94, 100, 168, 176, 179, 185

natural kinds 128–129, 139, 174, 179
neocortex 36–39, 44, 132, 172
nerves 9, 16–17, 27–28, **29**, 44, 54, 66, 70, 113, 176, 179, 185; cranial 28, **29**, 31, 44, 172, 179, 185; spinal 28–29, 44, 172, 184
neural network 106, 110, 180
neurodegeneration 42, 50, 75–76, 91, 94, 100, 111, 144, 153, 160, 170, 180
neurodiversity 142, **143**, 144, 160, 180
neuroendocrinology 174, 180
neurogenesis 22, 86, 92, 180

neuroimmunology 50, 177, 181
neuron(s) 7, 21–22, 26, 91, 105, 119–120, 163–165; afferent and efferent 23–24, 27, **29**; definition 181; development 85–87, 100; interneurons 38, 43–44; memory 106–107; mirror 89, 178; neuronal communication *see* synaptic transmission; preganglionic and postganglionic 27–28; sleep 97; stimulating and recording 40–41; structure 17, *18*, 19–20, 78, 179
neuroplasticity *see* synaptic plasticity
neurosis 147, 182
neurotransmission *see* synaptic transmission
neurotransmitter(s) *19*, 62–67, 131, 174, 182; acetylcholine 28, 63–66, 75, **77**, 106; dopamine 33–34, 64–65, 67, 73, 75–76, **77**, 97, 106, 134–136, 140, 155; glutamate 64–65, 67, 106, 168; histamine 49, 64, 76, **77**, 97; hypocretin/orexin 97–98; monoamine 32, 64, 76, 97, 169, 179; noradrenaline 28, 63–65, 76, **77**, 78, 97, 106; serotonin 64, 76–77, **77**, 97, 106; sleep 97
neurotoxic 67
niche construction 12–13, 162, 181
noncoding DNA 83
non-verbal communication 137
norepinephrine *see* noradrenaline
Nussbaum, Martha 129, 131

obesity 52, 54, 90, 158
Occam's Razor 10, 27, 181
occipital lobe *37*, 38, *39*, 86, 114
Ockham, William of 10
O'Keefe, John 107
Olds, James 133
optogenetics 40, 67–68, 179
Orbach, Susie 53
osmoregulation 54–56, 181, 185

Papez, James 36, 132
parasympathetic nervous system 27–28, 44, 180
Parkinson's disease, Parkinsonism 34, 50, 145, **151**, 153, 180; basal ganglia 170; cognition 153; inflammation 50; genes 94–95, 165; treatment 42, 66–67, 75, **77**, 155–157
Pavlovian 108, 110, 178
Pavlov, Ivan Petrovich 7, 73, 108
pedunculopontine tegmental nucleus 34
perception **26**, 38, 44, 59, 88, 112–115, 119–120, 128–129, 138, **149**, 163, 169, 175, 183
peripheral nervous system (PNS) 16, 27–30, 44, 46, 54, 57–58, 66, 172, 179–180; development 17; emotions 131, 136, 139–140
Pessoa, Luis 119
phrenology 6, 43, 123, 181
physicalism 11, 182, 184
Plutchik, Robert 127
positron emission tomography (PET) 42
primary motivated behaviours 57, 59, 130, 174, 179, 182, 185
prosopagnosia 134, 175
Proust, Marcel 114
psychopathology 130, 141–143, 182, 185; causes 144–147; diagnosis 147–152; neurological 152–154; treatments 155–160
psychosis 144–145, 147, 156, 158, 182

race 80
receptors 19, 42, 50–51, 57, 62–63, 65–67, 77–78, 106, 113, 130, 182; autoreceptors 63, 66; baroreceptors 56
reductionism/reductionist 10–13, 182, 184
reflex/reflexes 25, **26**, 27, 29, 33, 46, 51, 117

Reid, Thomas 114
Rexed, Bror *24*, 25, **26**, 43
Rice, Laura 160
Robbins, Trevor 69
Robinson, Terry 73
Russell, James 127
Ryle, Gilbert 124

Sacks, Oliver 75
Sapir, Edward 127
Schacter, Stanley 123
schizophrenia 31, 34, **77–78**, 87, 144–145, 147, **148**, 155, 183
Schwann, Theodore 7, 101
Schwann cells 176, 179
sensation 11, 15, 41, 58, 112–115, 119–120, 133, 159, 183; exteroceptive 46, 58, 112, 184; interoceptive 46, 58, 112, 184; senses **29**, 105, 112–114, 120, 184
Shallice, Tim 117–118
Singer, Jerome 123
Singer, Judy 142, 144
Sinha, Pawan 115
Skinner, Burrhus Frederic 8, 40, 69, 108–109, 133, 158
sleep 32, 50, 71, 95, 100, 104, 111, **150**, 153, 164, 169–170, 184; brain mechanisms 97; disorders 97–99, 177; function 99; structure 95, **96**, 185
somatosensory 25
specific hunger 130
spinal cord 9, 16–17, 22–23, *24*, 25, **26**, **29**, 30–31, *32*, 33–34, *37*, 38, 44, 100, 117, 130, 169, 171, 179–180
stem cells 20, 85, 100
stroke 55, 74, 76, 144, 154
substance abuse 70, 79, 93, 98, 118, 148, 177, 184
supervisory attentional system 117–118
sympathetic nervous system 27–28, 44, 180

synapse 17–18, *19*, 21, 48, 61–66, 93, 116, 135, 180, 184; monosynaptic 26; polysynaptic 26; synaptogenesis *see* neurogenesis
synaptic 26, 120, 163; cleft 18, *19*, 62, 67, 184; connections 20, 22, 66, 88, 103, 106–108, 111, 142; plasticity 94, 106, 181; pruning 22, 86–87, 91, 145; strengthening 87, 92, 106–107, 110, 184; transmission 17, *19*, 61–68, 78, 181

tabula rasa 4
temperature **26**, 34, 54, 57, 174, 182
temporal lobe 30, 35, *37*, 38, *39*, 114, 140, 168; emotion 133–135; language 138, 171; memory 93
thalamus 31, *32*, 34–36, *39*, 44, 99, 132, 145, 181, 185; attention 116; emotion 123; sensation 113–114, 120, 130; sleep 97; speech 138; thalamotomy 156
therapy 75–76, 160–161; behavior 158, 170; cognitive behavioral 159, 172; ECT 156; emotion focused 159; immunotherapy 75–76, 79, 155; non-invasive therapy 152; radiotherapy 93; rational emotive 159; replacement therapy 75
thermogenesis 54
thermoregulation *see* temperature
thirst 48, 55–56, 129–130, 139, 177, 181
Thorndike, Edward 7
Tizard, Jack 144
transcranial direct current stimulation (tDCS) 42
transcranial magnetic stimulation (TMS) 42, 156–157
Tversky, Amos 118

vagus nerve **29**, 46, 52–53, 56, 112, 185; vagal nerve stimulation 158

ventral stream *see* dorsal stream
ventricles (brain) 16, 22, 92, 145, 171, 185
ventricles (heart) 47, 48, 185
Vesalius, Andreas 4–5

water balance 54–56, 58, 129, 176, 181, 185
white matter 5, *24*, 25, *32*, 33, 37, 42, 44, 145, 172, 179, 185; aging 93; biopsychological functions 30–31; neurodegeneration 153
Whorf, Benjamin 127
Wild Boy of Aveyron 5
Willis, Thomas 5
Winston, Patrick 30
Wundt, Wilhelm 7, 127

Zuberbühler, Klaus 136

For Product Safety Concerns and Information please contact our EU representative GPSR@taylorandfrancis.com Taylor & Francis Verlag GmbH, Kaufingerstraße 24, 80331 München, Germany

Printed and bound by CPI Group (UK) Ltd, Croydon, CR0 4YY
08/06/2025
01897002-0002